Table of Problem-Solving Strategies

Note for users of the five-volume edition:
Volume 1 (pp. 1–481) includes chapters 1–15.
Volume 2 (pp. 482–607) includes chapters 16–19.
Volume 3 (pp. 608–779) includes chapters 20–24.
Volume 4 (pp. 780–1194) includes chapters 25–36.
Volume 5 (pp. 1148–1383) includes chapters 36–42.

Chapters 37–42 are not in the Standard Edition.

ActivPhysics™ OnLine Activities

 www.aw-bc.com/knight

Physics for Scientists and Engineers

Volume 3

A Strategic Approach

Randall D. Knight

California Polytechnic State University, San Luis Obispo

PEARSON

Addison
Wesley

San Francisco   Boston   New York
Cape Town   Hong Kong   London   Madrid   Mexico City
Montreal   Munich   Paris   Singapore   Sydney   Tokyo   Toronto

Executive Editor:	Adam Black, Ph.D.
Development Editor:	Alice Houston, Ph.D.
Project Manager:	Laura Kenney Editorial & Production Services
Associate Editor:	Liana Allday
Media Producer:	Claire Masson
Marketing Manager:	Christy Lawrence
Market Development:	Susan Winslow
Manufacturing Supervisor:	Vivian McDougal
Art Director:	Blakely Kim
Production Service:	Thompson Steele, Inc.
Text Design:	Mark Ong, Side by Side Studios
Cover Design:	Yvo Riezebos Design
Illustrations:	Precision Graphics
Photo Research:	Cypress Integrated Systems
Cover Printer:	Phoenix Color Corporation
Printer and Binder:	R. R. Donnelley & Sons
Cover Image:	Rainbow/PictureQuest
Credits:	see page C–1

Library of Congress Cataloging-in-Publication Data

Knight, Randall Dewey.
    Physics for scientists and engineers : a strategic approach / Randall D. Knight.
      p. cm.
    Includes index.
    ISBN 0-8053-8960-1 (extended ed. with MasteringPhysics)
     1. Physics   I. Title.

  QC23.2.K65 2004
  530--dc22

                                                 2003062809

ISBN 0-8053-8970-9 Volume 3 with MasteringPhysics
ISBN 0-8053-9014-6 Volume 3 without MasteringPhysics

5 6 7 8 9 10—DOW—06
www.aw-bc.com

Brief Contents

About the Author

Randy Knight has taught introductory physics for over 20 years at Ohio State University and California Polytechnic University, where he is currently Professor of Physics. Professor Knight received a bachelor's degree in physics from Washington University in St. Louis and a Ph.D. in physics from the University of California, Berkeley. He was a post-doctoral fellow at the Harvard-Smithsonian Center for Astrophysics before joining the faculty at Ohio State University. It was at Ohio State that he began to learn about the research in physics education that, many years later, led to this book.

Professor Knight's research interests are in the field of lasers and spectroscopy, and he has published over 25 research papers. He recently led the effort to establish an environmental studies program at Cal Poly, where, in addition to teaching introductory physics, he also teaches classes on energy, oceanography, and environmental issues. When he's not in the classroom or in front of a computer, you can find Randy hiking, sea kayaking, playing the piano, or spending time with his wife Sally and their seven cats.

Preface to the Instructor

In 1997 we published *Physics: A Contemporary Perspective.* This was the first comprehensive, calculus-based textbook to make extensive use of results from physics education research. The development and testing that led to this book had been partially funded by the National Science Foundation. In the preface we noted that it was a "work in progress" and that we very much wanted to hear from users—both instructors and students—to help us shape the book into a final form.

And hear from you we did! We received feedback and reviews from roughly 150 professors and, especially important, 4500 of their students. This textbook, the newly titled *Physics for Scientists and Engineers: A Strategic Approach*, is the result of synthesizing that feedback and using it to produce a book that we hope is uniquely tuned to helping today's students succeed. It is the first introductory textbook built from the ground up on research into how students can more effectively learn physics.

Objectives

My primary goals in writing *Physics for Scientists and Engineers: A Strategic Approach* have been:

- To produce a textbook that is more focused and coherent, less encyclopedic.
- To move key results from physics education research into the classroom in a way that allows instructors to use a range of teaching styles.
- To provide a balance of quantitative reasoning and conceptual understanding, with special attention to concepts known to cause student difficulties.
- To develop students' problem-solving skills in a systematic manner.
- To support an active-learning environment.

These goals and the rationale behind them are discussed at length in my small paperback book, *Five Easy Lessons: Strategies for Successful Physics Teaching* (Addison Wesley, 2002). Please request a copy from your local Addison Wesley sales representative if it would be of interest to you (ISBN 0-8053-8702-1).

Textbook Organization

The 42-chapter extended edition (ISBN 0-8053-8685-8) of *Physics for Scientists and Engineers* is intended for use in a three-semester course. Most of the 36-chapter standard edition (ISBN 0-8053-8982-2), ending with relativity, can be covered in two semesters, but the judicious omission of a few chapters will avoid rushing through the material and give students more time to develop their knowledge and skills.

There's a growing sentiment that quantum physics is quickly becoming the province of engineers, not just scientists, and that even a two–semester course should include a reasonable introduction to quantum ideas. The *Instructor's Guide* outlines a couple of routes through the book that allow most of the quantum physics chapters to be reached in two semesters. I've written the book with the hope that an increasing number of instructors will choose one of these routes.

- **Extended edition,** with modern physics (ISBN 0-8053-8685-8): chapters 1–42.
- **Standard edition** (ISBN 0-8053-8982-2): chapters 1–36.
- **Volume 1** (ISBN 0-8053-8963-6) covers mechanics: chapters 1–15.
- **Volume 2** (ISBN 0-8053-8966-0) covers thermodynamics: chapters 16–19.
- **Volume 3** (ISBN 0-8053-8969-5) covers waves and optics: chapters 20–24.
- **Volume 4** (ISBN 0-8053-8972-5) covers electricity and magnetism, plus relativity: chapters 25–36.
- **Volume 5** (ISBN 0-8053-8975-X) covers relativity and quantum physics: chapters 36–42.
- **Volumes 1–5** boxed set (ISBN 0-8053-8978-4).

The full textbook is divided into seven parts: Part I: *Newton's Laws*, Part II: *Conservation Laws*, Part III: *Applications of Newtonian Mechanics*, Part IV: *Thermodynamics*, Part V: *Waves and Optics*, Part VI: *Electricity and Magnetism*, and Part VII: *Relativity and Quantum Mechanics*. Although I recommend covering the parts in this order (see below), doing so is by no means essential. Each topic is self-contained, and Parts III–VI can be rearranged to suit an instructor's needs. To facilitate a reordering of topics, the full text is available in the five individual volumes listed in the margin.

Organization Rationale: Thermodynamics is placed before waves because it is a continuation of ideas from mechanics. The key idea in thermodynamics is energy, and moving from mechanics into thermodynamics allows the uninterrupted development of this important idea. Further, waves introduce students to functions of two variables, and the mathematics of waves is more akin to electricity and magnetism than to mechanics. Thus moving from waves to fields to quantum physics provides a gradual transition of ideas and skills.

The purpose of placing optics with waves is to provide a coherent presentation of wave physics, one of the two pillars of classical physics. Optics as it is presented in introductory physics makes no use of the properties of electromagnetic fields. There's little reason other than historical tradition to delay optics until after E&M. The documented difficulties that students have with optics are difficulties with waves, not difficulties with electricity and magnetism. However, the optics chapters are easily deferred until the end of Part VI for instructors who prefer that ordering of topics.

More Effective Problem-Solving Instruction

Careful and systematic instruction is provided on all aspects of problem solving. Some of the features that support this approach are described here, and more details are provided in the *Instructor's Guide*.

- An emphasis on using *multiple representations*—descriptions in words, pictures, graphs, and mathematics—to look at a problem from many perspectives.
- The explicit use of *models*, such as the particle model, the wave model, and the field model, to help students recognize and isolate the essential features of a physical process.
- TACTICS BOXES for the development of particular skills, such as drawing a free-body diagram or using Lenz's law. Tactics Box steps are explicitly illustrated in subsequent worked examples, and these are often the starting point of a full problem-solving strategy.

TACTICS BOX 4.3 **Drawing a free-body diagram**

❶ **Identify all forces acting on the object.** This step was described in Tactics Box 4.2.

❷ **Draw a coordinate system.** Use the axes defined in your pictorial representation. If those axes are tilted, for motion along an incline, then the axes of the free-body diagram should be similarly tilted.

❸ **Represent the object as a dot at the origin of the coordinate axes.** This is the particle model.

❹ **Draw vectors representing each of the identified forces.** This was described in Tactics Box 4.1. Be sure to label each force vector.

❺ **Draw and label the *net force* vector $\vec{F}_{net}$.** Draw this vector beside the diagram, not on the particle. Or, if appropriate, write $\vec{F}_{net} = \vec{0}$. Then check that $\vec{F}_{net}$ points in the same direction as the acceleration vector $\vec{a}$ on your motion diagram.

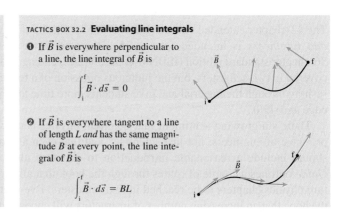

TACTICS BOX 32.2 **Evaluating line integrals**

❶ If $\vec{B}$ is everywhere perpendicular to a line, the line integral of $\vec{B}$ is

$$\int_i^f \vec{B} \cdot d\vec{s} = 0$$

❷ If $\vec{B}$ is everywhere tangent to a line of length L *and* has the same magnitude B at every point, the line integral of $\vec{B}$ is

$$\int_i^f \vec{B} \cdot d\vec{s} = BL$$

- **PROBLEM-SOLVING STRATEGIES** that help students develop confidence and more proficient problem-solving skills through the use of a consistent four-step approach: **MODEL, VISUALIZE, SOLVE, ASSESS**. Strategies are provided for each broad class of problems, such as dynamics problems or problems involving electromagnetic induction. The (MP) icon directs students to the specially developed *Skill Builder* tutorial problems in MasteringPhysics™ (see page xi), where they can interactively work through each of these strategies online.

- Worked **EXAMPLES** that illustrate good problem-solving practices through the consistent use of the four-step problem-solving approach and, where appropriate, the Tactics Box steps. The worked examples are often very detailed and carefully lead the student step by step through the *reasoning* behind the solution, not just through the numerical calculations. Steps that are often implicit or omitted in other textbooks, because they seem so obvious to experts, are explicitly discussed since research has shown these are often the points where students become confused.

- **NOTE ▶** Paragraphs within worked examples caution against common mistakes and point out useful tips for tackling problems.

- The *Student Workbook* (see page xi), a unique component of this text, bridges the gap between worked examples and end-of-chapter problems. It provides qualitative problems and exercises that focus on developing the skills and conceptual understanding necessary to solve problems with confidence.

- Approximately 3000 original and diverse *end-of-chapter problems* have been carefully crafted to exercise and test the full range of qualitative and quantitative problem-solving skills. *Exercises*, which are keyed to specific sections, allow students to practice basic skills and computations. *Problems* require a better understanding of the material and often draw upon multiple representations of knowledge. *Challenge Problems* are more likely to use calculus, utilize ideas from more than one chapter, and sometimes lead students to explore topics that weren't explicitly covered in the chapter.

(MP) **PROBLEM-SOLVING STRATEGY 5.2** **Dynamics problems**

MODEL Make simplifying assumptions.

VISUALIZE

Pictorial representation. Show important points in the motion with a sketch, establish a coordinate system, define symbols, and identify what the problem is trying to find. This is the process of translating words to symbols.

Physical representation. Use a motion diagram to determine the object's acceleration vector $\vec{a}$. Then identify all forces acting on the object and show them on a free-body diagram.

It's OK to go back and forth between these two steps as you visualize the situation.

SOLVE The mathematical representation is based on Newton's second law

$$\vec{F}_{net} = \sum_i \vec{F}_i = m\vec{a}$$

The vector sum of the forces is found directly from the free-body diagram. Depending on the problem, either

- Solve for the acceleration, then use kinematics to find velocities and positions, or
- Use kinematics to determine the acceleration, then solve for unknown forces.

ASSESS Check that your result has the correct units, is reasonable, and answers the question.

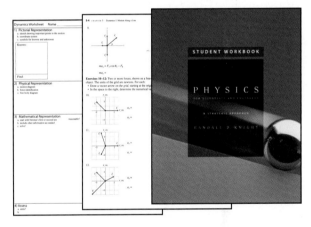

Proven Features to Promote Deeper Understanding

Research has shown that many students taking calculus-based physics arrive with a wealth of misconceptions and subsequently struggle to develop a coherent understanding of the subject. Using a number of unique, reinforcing techniques, this book tackles these issues head-on to enable students to build a solid foundation of understanding.

- A *concrete-to-abstract* approach introduces new concepts through observations about the real world and everyday experience. Step by step, the text then builds up the concepts and principles needed by a theory that will make sense of the observations and make new, testable predictions. This inductive approach better matches how students learn, and it reinforces how physics—and science in general—operates.

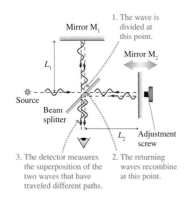

Annotated **FIGURE** showing the operation of the Michelson interferometer.

■ **STOP TO THINK** questions embedded in each chapter allow students to assess whether they've understood the main idea of a section. The *Stop to Think* questions, which include concept questions, ratio reasoning, and ranking tasks, are primarily derived from physics education research.

■ **NOTE ▶** paragraphs draw attention to common misconceptions, clarify possible confusions in terminology and notation, and provide important links to previous topics.

■ Unique *annotated figures*, based on research into visual learning modes, make the artwork a teaching tool on a par with the written text. Commentary in blue—the "instructor's voice"—helps students "read" the figure. Students "learn by viewing" how to interpret a graph, how to translate between multiple representations, how to grasp a difficult concept through a visual analogy, and many other important skills.

■ The learning goals and links that begin each chapter outline what the student needs to remember from previous chapters and what to focus on in the chapter ahead.

    ▶ **Looking Ahead** lists key concepts and skills the student will learn in the coming chapter.

    ◀ **Looking Back** suggests important topics students should review from previous chapters.

■ Unique schematic *Chapter Summaries* help students organize their knowledge in an expert-like hierarchy, from general principles (top) to applications (bottom). Side-by-side pictorial, graphical, textual, and mathematical representations are used to help students with different learning styles and enable them to better translate between these key representations.

■ *Part Overviews and Summaries* provide a global framework for the student's learning. Each part begins with an overview of the chapters ahead. It then concludes with a broad summary to help students draw connections between the concepts presented in that set of chapters. **KNOWLEDGE STRUCTURE** tables in the part summaries, similar to the chapter summaries, help students see a forest rather than dozens of individual trees.

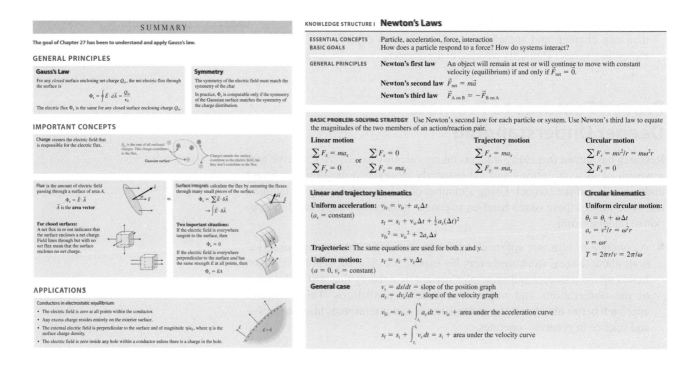

The Student Workbook

A key component of *Physics for Scientists and Engineers: A Strategic Approach* is the accompanying *Student Workbook*. The workbook bridges the gap between textbook and homework problems by providing students the opportunity to learn and practice skills prior to using those skills in quantitative end-of-chapter problems, much as a musician practices technique separately from performance pieces. The workbook exercises, which are keyed to each section of the textbook, focus on developing specific skills, ranging from identifying forces and drawing free-body diagrams to interpreting wave functions.

The workbook exercises, which are generally qualitative and/or graphical, draw heavily upon the physics education research literature. The exercises deal with issues known to cause student difficulties and employ techniques that have proven to be effective at overcoming those difficulties. The workbook exercises can be used in-class as part of an active-learning teaching strategy, in recitation sections, or as assigned homework. More information about effective use of the *Student Workbook* can be found in the *Instructor's Guide*.

Available versions: Extended (ISBN 0-8053-8961-X), Standard (ISBN 0-8053-8984-9), Volume 1 (ISBN 0-8053-8965-2), Volume 2 (ISBN 0-8053-8968-7), Volume 3 (ISBN 0-8053-8971-7), Volume 4 (ISBN 0-8053-8974-1), and Volume 5 (ISBN 0-8053-8977-6).

Instructor Supplements

- The **Instructor's Guide for Physics for Scientists and Engineers** (ISBN 0-8053-8985-7) offers detailed comments and suggested teaching ideas for every chapter, an extensive review of what has been learned from physics education research, and guidelines for using active-learning techniques in your classroom.

- The **Instructor's Solutions Manuals**, **Chapters 1–19** (ISBN 0-8053-8986-5), and **Chapters 20–42** (ISBN 0-8053-8989-X), written by Professors Pawan Kahol and Donald Foster, at Wichita State University, provide *complete* solutions to all the end-of-chapter problems. The solutions follow the four-step Model/Visualize/Solve/Assess procedure used in the *Problem-Solving Strategies* and all worked examples. Emphasis is placed on the reasoning behind the solution, rather than just the numerical manipulations. The full text of each solution is available as an editable Word document and as a pdf file on the *Instructor's Supplement CD-ROM* for your own use or for posting on your course website.

- The cross-platform **Instructor's Resource CD-ROMs** (ISBN 0-8053-8996-2) consists of the **Simulation and Image Presentation CD-ROM** and the **Instructor's Supplement CD-ROM**. The *Simulation and Image Presentation CD-ROM* provides a comprehensive library of more than 220 applets from *ActivPhysics OnLine*, as well as all the figures from the textbook (excluding photographs) in JPEG format. In addition, all the tables, chapter summaries, and knowledge structures are provided as JPEGs, and the Tactics Boxes, Problem-Solving Strategies, and key (boxed) equations are provided in editable Word format. The *Instructor's Supplement CD-ROM* provides editable Word versions and pdf files of the *Instructor's Guide* and the *Instructor's Solutions Manuals*. Complete *Student Workbook* solutions are also provided as pdf files.

- **MasteringPhysics**™ (www.masteringphysics.com) is a sophisticated, research-proven online tutorial and homework assignment system that provides students with individualized feedback and hints based on their input. It provides a comprehensive library of conceptual tutorials (including one for each

Problem-Solving Strategy in this textbook), multistep self-tutoring problems, and end-of-chapter problems from *Physics for Scientists and Engineers*. *MasteringPhysics*™ provides instructors with a fast and effective way to assign online homework assignments that comprise a range of problem types. The powerful post-assignment diagnostics allow instructors to assess the progress of their class as a whole or to quickly identify individual students' areas of difficulty.

- **ActivPhysics**™ **OnLine** (www.aw-bc.com/knight) provides a comprehensive library of more than 420 tried and tested *ActivPhysics* applets updated for web delivery using the latest online technologies. In addition, it provides a suite of highly regarded applet-based tutorials developed by education pioneers Professors Alan Van Heuvelen and Paul D'Alessandris. The *ActivPhysics* margin icon directs students to specific exercises that complement the textbook discussion.

  The online exercises are designed to encourage students to confront misconceptions, reason qualitatively about physical processes, experiment quantitatively, and learn to think critically. They cover all topics from mechanics to electricity and magnetism and from optics to modern physics. The highly acclaimed *ActivPhysics OnLine* companion workbooks help students work through complex concepts and understand them more clearly. More than 220 applets from the *ActivPhysics OnLine* library are also available on the *Simulation and Image Presentation CD-ROM*.

- The **Printed Test Bank** (ISBN 0-8053-8994-6) and cross-platform **Computerized Test Bank** (ISBN 0-8053-8995-4), prepared by Professor Benjamin Grinstein, at the University of California, San Diego, contain more than 1500 high-quality problems, with a range of multiple-choice, true/false, short-answer, and regular homework-type questions. In the computerized version, more than half of the questions have numerical values that can be randomly assigned for each student.

- The **Transparency Acetates** (ISBN 0-8053-8993-8) provide more than 200 key figures from *Physics for Scientists and Engineers* for classroom presentation.

Student Supplements

- The **Student Solutions Manuals Chapters 1–19** (ISBN 0-8053-8708-0) and **Chapters 20–42** (ISBN 0-8053-8998-9), written by Professors Pawan Kahol and Donald Foster at Wichita State University, provides *detailed* solutions to more than half of the odd-numbered end-of-chapter problems. The solutions follow the four-step Model/Visualize/Solve/Assess procedure used in the *Problem-Solving Strategies* and all worked examples.

- **MasteringPhysics**™ (www.masteringphysics.com) provides students with individualized online tutoring by responding to their wrong answers and providing hints for solving multistep problems. It gives them immediate and up-to-date assessment of their progress, and shows where they need to practice more.

- **ActivPhysics**™ **OnLine** (www.aw-bc.com/knight) provides students with a suite of highly regarded applet-based tutorials (see above). The accompanying workbooks help students work though complex concepts and understand them more clearly. The *ActivPhysics* margin icon directs students to specific exercises that complement the textbook discussion.

- **ActivPhysics OnLine Workbook Volume 1: Mechanics • Thermal Physics • Oscillations & Waves** (ISBN 0-8053-9060-X)

- **ActivPhysics OnLine Workbook Volume 2: Electricity & Magnetism • Optics • Modern Physics** (ISBN 0-8053-9061-8)

■ The **Addison-Wesley Tutor Center** (www.aw.com/tutorcenter) provides one-on-one tutoring via telephone, fax, email, or interactive website during evening hours and on weekends. Qualified college instructors answer questions and provide instruction for *Mastering Physics*™ and for the examples, exercises, and problems in *Physics for Scientists and Engineers*.

Acknowledgments

I have relied upon conversations with and, especially, the written publications of many members of the physics education community. Those who may recognize their influence include Arnold Arons, Uri Ganiel, Ibrahim Halloun, Richard Hake, David Hestenes, Leonard Jossem, Jill Larkin, Priscilla Laws, John Mallinckrodt, Lillian McDermott, Edward "Joe" Redish, Fred Reif, Rachel Scherr, Bruce Sherwood, David Sokoloff, Ronald Thornton, Sheila Tobias, and Alan Van Heuleven. John Rigden, founder and director of the Introductory University Physics Project, provided the impetus that got me started down this path. Early development of the materials was supported by the National Science Foundation as the *Physics for the Year 2000* project; their support is gratefully acknowledged.

I am grateful to Pawan Kahol and Don Foster for the difficult task of writing the *Instructor's Solutions Manuals*; to Jim Andrews and Susan Cable for writing the workbook answers; to Wayne Anderson, Jim Andrews, Dave Ettestad, Stuart Field, Robert Glosser, and Charlie Hibbard for their contributions to the end-of-chapter problems; and to my colleague Matt Moelter for many valuable contributions and suggestions.

I especially want to thank my editor Adam Black, development editor Alice Houston, editorial assistant Liana Allday, and all the other staff at Addison Wesley for their enthusiasm and hard work on this project. Project manager Laura Kenney, Carolyn Field and the team at Thompson Steele, Inc., copy editor Kevin Gleason, photo researcher Brian Donnelly, and page-layout artist Judy Maenle get much of the credit for making this complex project all come together. In addition to the reviewers and classroom testers listed below, who gave invaluable feedback, I am particularly grateful to Wendell Potter and Susan Cable for their close scrutiny of every word and figure.

Finally, I am endlessly grateful to my wife Sally for her love, encouragement, and patience, and to our many cats for their innate abilities to hold down piles of papers and to type qqqqqqqq whenever it was needed.

Randy Knight, September 2003
rknight@calpoly.edu

Reviewers and Classroom Testers

Gary B. Adams, *Arizona State University*
Wayne R. Anderson, *Sacramento City College*
James H. Andrews, *Youngstown State University*
David Balogh, *Fresno City College*
Dewayne Beery, *Buffalo State College*
Joseph Bellina, *Saint Mary's College*
James R. Benbrook, *University of Houston*
David Besson, *University of Kansas*

Randy Bohn, *University of Toledo*
Art Braundmeier, *University of Southern Illinois, Edwardsville*
Carl Bromberg, *Michigan State University*
Douglas Brown, *Cabrillo College*
Ronald Brown, *California Polytechnic State University, San Luis Obispo*
Mike Broyles, *Collin County Community College*

James Carolan, *University of British Columbia*
Michael Crescimanno, *Youngstown State University*
Wei Cui, *Purdue University*
Robert J. Culbertson, *Arizona State University*
Purna C. Das, *Purdue University North Central*
Dwain Desbien, *Estrella Mountain Community College*
John F. Devlin, *University of Michigan, Dearborn*
Alex Dickison, *Seminole Community College*
Chaden Djalali, *University of South Carolina*
Sandra Doty, *Denison University*
Miles J. Dresser, *Washington State University*
Charlotte Elster, *Ohio University*
Robert J. Endorf, *University of Cincinnati*
Tilahun Eneyew, *Embry-Riddle Aeronautical University*
F. Paul Esposito, *University of Cincinnati*
John Evans, *Lee University*
Michael R. Falvo, *University of North Carolina*
Abbas Faridi, *Orange Coast College*
Stuart Field, *Colorado State University*
Daniel Finley, *University of New Mexico*
Jane D. Flood, *Muhlenberg College*
Thomas Furtak, *Colorado School of Mines*
Richard Gass, *University of Cincinnati*
J. David Gavenda, *University of Texas, Austin*
Stuart Gazes, *University of Chicago*
Katherine M. Gietzen, *Southwest Missouri State University*
Robert Glosser, *University of Texas, Dallas*
William Golightly, *University of California, Berkeley*
Paul Gresser, *University of Maryland*
C. Frank Griffin, *University of Akron*
John B. Gruber, *San Jose State University*
Randy Harris, *University of California, Davis*
Stephen Haas, *University of Southern California*
Nicole Herbots, *Arizona State University*
Scott Hildreth, *Chabot College*
David Hobbs, *South Plains College*
Laurent Hodges, *Iowa State University*
John L. Hubisz, *North Carolina State University*
George Igo, *University of California, Los Angeles*
Bob Jacobsen, *University of California, Berkeley*
Rong-Sheng Jin, *Florida Institute of Technology*
Marty Johnston, *University of St. Thomas*
Stanley T. Jones, *University of Alabama*
Darrell Judge, *University of Southern California*
Pawan Kahol, *Wichita State University*
Teruki Kamon, *Texas A&M University*
Richard Karas, *California State University, San Marcos*
Deborah Katz, *U.S. Naval Academy*
Miron Kaufman, *Cleveland State University*
M. Kotlarchyk, *Rochester Institute of Technology*
Cagliyan Kurdak, *University of Michigan*
Fred Krauss, *Delta College*
H. Sarma Lakkaraju, *San Jose State University*

Darrell R. Lamm, *Georgia Institute of Technology*
Robert LaMontagne, *Providence College*
Alessandra Lanzara, *University of California, Berkeley*
Sen-Ben Liao, *Massachusetts Institute of Technology*
Dean Livelybrooks, *University of Oregon*
Chun-Min Lo, *University of South Florida*
Richard McCorkle, *University of Rhode Island*
James McGuire, *Tulane University*
Theresa Moreau, *Amherst College*
Gary Morris, *Rice University*
Michael A. Morrison, *University of Oklahoma*
Richard Mowat, *North Carolina State University*
Taha Mzoughi, *Mississippi State University*
Vaman M. Naik, *University of Michigan, Dearborn*
Craig Ogilvie, *Iowa State University*
Martin Okafor, *Georgia Perimeter College*
Benedict Y. Oh, *University of Wisconsin*
Georgia Papaefthymiou, *Villanova University*
Peggy Perozzo, *Mary Baldwin College*
Brian K. Pickett, *Purdue University, Calumet*
Joe Pifer, *Rutgers University*
Dale Pleticha, *Gordon College*
Robert Pompi, *SUNY-Binghamton*
David Potter, *Austin Community College*
Chandra Prayaga, *University of West Florida*
Didarul Qadir, *Central Michigan University*
Michael Read, *College of the Siskiyous*
Michael Rodman, *Spokane Falls Community College*
Sharon Rosell, *Central Washington University*
Anthony Russo, *Okaloosa-Walton Community College*
Otto F. Sankey, *Arizona State University*
Rachel E. Scherr, *University of Maryland*
Bruce Schumm, *University of California, Santa Cruz*
Douglas Sherman, *San Jose State University*
Elizabeth H. Simmons, *Boston University*
Alan Slavin, *Trent College*
William Smith, *Boise State University*
Paul Sokol, *Pennsylvania State University*
Chris Sorensen, *Kansas State University*
Anna and Ivan Stern, *AW Tutor Center*
Michael Strauss, *University of Oklahoma*
Arthur Viescas, *Pennsylvania State University*
Chris Vuille, *Embry-Riddle Aeronautical University*
Ernst D. Von Meerwall, *University of Akron*
Robert Webb, *Texas A&M University*
Zodiac Webster, *California State University, San Bernardino*
Robert Weidman, *Michigan Technical University*
Jeff Allen Winger, *Mississippi State University*
Ronald Zammit, *California Polytechnic State University, San Luis Obispo*
Darin T. Zimmerman, *Pennsylvania State University, Altoona*

Preface to the Student

From Me to You

The most incomprehensible thing about the universe is that it is comprehensible.
  —Albert Einstein

The day I went into physics class it was death.
  —Sylvia Plath, *The Bell Jar*

Let's have a little chat before we start. A rather one-sided chat, admittedly, because you can't respond, but that's OK. I've heard from many of your fellow students over the years, so I have a pretty good idea of what's on your mind.

What's your reaction to taking physics? Fear and loathing? Uncertainty? Excitement? All of the above? Let's face it, physics has a bit of an image problem on campus. You've probably heard that it's difficult, maybe downright impossible unless you're an Einstein. Things that you've heard, your experiences in other science courses, and many other factors all color your *expectations* about what this course is going to be like.

It's true that there are many new ideas to be learned in physics and that the course, like college courses in general, is going to be much faster paced than science courses you had in high school. I think it's fair to say that it will be an *intense* course. But we can avoid many potential problems and difficulties if we can establish, here at the beginning, what this course is about and what is expected of you—and of me!

Just what is physics, anyway? Physics is a way of thinking about the physical aspects of nature. Physics is not better than art or biology or poetry or religion, which are also ways to think about nature; it's simply different. One of the things this course will emphasize is that physics is a human endeavor. The information content of this book was not found in a cave or conveyed to us by aliens; it was discovered by real people engaged in a struggle with real issues. I hope to convey to you something of the history and the process by which we have come to accept the principles that form the foundation of today's science and engineering.

You might be surprised to hear that physics is not about "facts." Oh, not that facts are unimportant, but physics is far more focused on discovering *relationships* that exist between facts and *patterns* that exist in nature than on learning facts for their own sake. As a consequence, there's not a lot of memorization when you study physics. Some—there are still definitions and equations to learn—but less than in many other courses. Our emphasis, instead, will be on thinking and reasoning. This is important to factor into your expectations for the course.

Perhaps most important of all, *physics is not math!* Physics is much broader. We're going to look for patterns and relationships in nature, develop the logic that relates different ideas, and search for the reasons *why* things happen as they do. In doing so, we're going to stress qualitative reasoning, pictorial and graphical reasoning, and reasoning by analogy. And yes, we will use math, but it's just one tool among many.

It will save you much frustration if you're aware of this physics–math distinction up front. Many of you, I know, want to find a formula and plug numbers into it—that is, to do a math problem. Maybe that's what you learned in high school science courses, but it is *not* what this course expects of you. We'll certainly do

(a) X-ray diffraction pattern

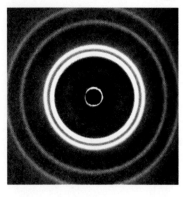

(b) Electron diffraction pattern

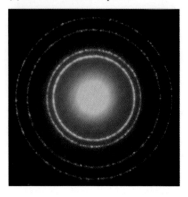

many calculations, but the specific numbers are usually the last and least important step in the analysis.

Physics is about recognizing patterns. The top photograph is an x-ray diffraction pattern that shows how a collimated beam of x rays spreads out after passing through a crystal. The bottom photograph shows what happens when a collimated beam of electrons is shot through the same crystal. What does the obvious similarity in these two photographs tell us about the nature of light and about the nature of matter?

As you study, you'll sometimes be baffled, puzzled, and confused. That's perfectly normal and to be expected. Making mistakes is OK too *if* you're willing to learn from the experience. No one is born knowing how to do physics any more than he or she is born knowing how to play the piano or shoot basketballs. The ability to do physics comes from practice, repetition, and struggling with the ideas until you "own" them and can apply them yourself in new situations. There's no way to make learning effortless, at least for anything worth learning, so expect to have some difficult moments ahead.

But also expect to have some moments of excitement at the joy of discovery. There will be instants at which the pieces suddenly click into place and you *know* that you understand a difficult idea. There will be times when you'll surprise yourself by successfully working a difficult problem that you didn't think you could solve. My hope, as an author, is that the excitement and sense of adventure will far outweigh the difficulties and frustrations.

Many of you, I suspect, would like to know the "best" way to study for this course. There is no best way. People are too different, and what works for one student works less effectively for another. But I do want to stress that *reading the text* is vitally important. Class time will be used to clarify difficulties and to develop tools for using the knowledge, but your instructor will *not* use class time simply to repeat information in the text. The basic knowledge for this course is written down within these pages, and the *number one expectation* is that you will read carefully and thoroughly to find and learn that knowledge.

Despite there being no best way to study, I will suggest *one* way that is successful for many students. It consists of the following four steps:

1. **Read each chapter *before* it is discussed in class.** I cannot stress too highly how important this step is. Class attendance is largely ineffective if you have not prepared. When you first read a chapter, focus on learning new vocabulary, definitions, and notation. There's a list of terms and notations at the end of each chapter. Learn them! You won't understand what's being discussed or how the ideas are being used if you don't know what the terms and symbols mean.

2. **Participate actively in class.** Take notes, ask and answer questions, take part in discussion groups. There is ample scientific evidence that *active participation* is far more effective for learning science than is passive listening.

3. **After class, go back for a *careful* rereading of the chapter.** In your second reading, pay closer attention to the details and the worked examples. Look for the *logic* behind each example (and I've tried to help make this clear), not just at what formula is being used. Do the *Student Workbook* exercises for each section as you finish your reading of it.

4. **Finally, apply what you have learned to the homework problems at the end of each chapter.** I strongly encourage you to form a study group with two or three classmates. There's good evidence that students who study regularly with a group do better than the rugged individualists who try to go it alone.

Did someone mention a workbook? The companion *Student Workbook* is a vital part of this course. It contains questions and exercises that ask you to reason *qualitatively*, to use graphical information, and to give explanations. It is through these exercises that you will learn what the concepts mean and will practice the reasoning skills appropriate to the chapter. You will then have acquired the baseline knowledge that you need *before* turning to the end-of-chapter homework problems. In sports or in music, you would never think of performing before you practice, so why would you want to do so in physics? The workbook is where you practice and work on basic skills.

Many of you, I know, would like to go straight to the homework problems and then thumb through the text looking for a formula that seems like it will work. That approach will not succeed in this course, and it's guaranteed to make you frustrated and discouraged. Very few homework problems are "plug and chug" problems where you simply put numbers into a formula. To work the homework problems successfully, you need a better study strategy—either that outlined above or your own—that helps you learn the concepts and the relationships between the ideas. Many of the chapters in this book have Problem-Solving Strategies to help you develop effective problem-solving skills.

A traditional guideline in college is to study two hours outside of class for every hour spent in class, and this text is designed with that expectation. Of course, two hours is an average. Some chapters are fairly straightforward and will go quickly. Others likely will require much more than two study hours per class hour.

Now that you know more about what is expected of you, what can you expect of me? That's a little trickier, because the book is already written! Nonetheless, it was prepared on the basis of what I think my students throughout the years have expected—and wanted—from their physics textbook.

You should know that these course materials—the text and the workbook—are based upon extensive research about how students learn physics and the challenges they face. The effectiveness of many of the exercises has been demonstrated through extensive class testing. I've written the book in an informal style that I hope you will find appealing and that will encourage you to do the reading. And finally, I have endeavored to make clear not only that physics, as a technical body of knowledge, is relevant to your profession but also that physics is an exciting adventure of the human mind.

I hope you'll enjoy the time we're going to spend together.

Detailed Contents

Volume 1 contains chapters 1–15; Volume 2 contains chapters 16-19; Volume 3 contains chapters 20–24; Volume 4 contains chapters 25–36; Volume 5 contains chapters 36–42.

If violent storms drive waves across the ocean of planet Kamino at a speed of 75 m/s, what is the wavelength of an ocean wave whose period is 10 s?

Waves and Optics

Beyond the Particle Model

Parts I–IV of this text have been about the physics of particles. You saw that macroscopic systems, such as solids and gases, can be thought of as systems of particles. A *particle* is one of the two fundamental models of physics. The other, to which we now turn our attention, is a *wave*.

Waves are ubiquitous in nature. Familiar examples of waves include

- Ripples on a pond.
- Surf crashing on a beach.
- The swaying ground of an earthquake.
- A vibrating guitar string.
- The sweet sound of a flute.
- The colors of the rainbow.
- A laser beam shooting through space.

The physics of waves is the subject of Part V, the next stage of our journey. Despite the great diversity of types and sources of waves, there is a single, elegant physical theory that is capable of describing them all. Our exploration of wave phenomena will call upon water waves, sound waves, and light waves for examples, but we want to emphasize the unity and coherence of the ideas that are common to *all* types of waves.

We will start with waves that travel outward through some medium, like the ripples that spread out from a pebble hitting a pool of water or the sounds that emanate from a loudspeaker. These are called *traveling waves*. After arriving at an understanding of a single traveling wave, we will see what happens when several traveling waves are combined. This investigation will lead us to *standing waves*, which are essential for understanding music and lasers, and to the phenomenon called *interference*, one of the most important defining characteristics of waves.

Two chapters will be devoted to light and optics, perhaps the most important application of waves. Light is an elusive subject, and we will end up developing three different models of light:

- The *wave model*, which explains the interference and diffraction of light.
- The *ray model*, which provides a framework for understanding how lenses and mirrors form images.
- The *photon model*, which will bring us to the foothills of quantum physics.

Although light is an electromagnetic wave, these chapters depend on nothing more than the "waviness" of light waves for your understanding. You can study these chapters either before or after your study of electricity and magnetism in

Part VI. The electromagnetic aspects of light waves will be taken up in Chapter 34.

Finally, waves will give us new information about atoms and their properties. Much of the light and color in our environment is due to the emission and absorption of light by atoms and molecules. The atoms of each element in the periodic table emit a unique "fingerprint" of light, a fingerprint that can be read with optical instruments and that provides clues about the structure of the atoms. The search for a theory that could successfully decipher these clues ultimately led to the development of quantum physics at the beginning of the 20th century. Part V will conclude with an initial look at the connection between atoms and light. We will then return to this important topic in Part VII.

Waves and Particles

Our journey through Part V will lead us to a closer examination of the relationship between waves and particles. The classical physics of the 18th and 19th centuries made a fundamental distinction between waves and particles. There was a well-defined wave-particle dichotomy, an either-or situation, and each of the objects in nature could be characterized as one or the other. Planets, projectiles, and atoms were clearly particles, or systems of particles, while sound and light were waves. That is not to say that particles (air molecules) are not relevant to sound waves, but sound itself is a collective, wave-like behavior of the air. Sound does not follow a parabolic trajectory as a particle would.

Table OV.V.1 lists some of the basic, defining characteristics of the wave and particle models. The idea that a particle exists at a specific location is particularly important. It has a position coordinate x. You can put your finger on it. It is *here*. A wave, by contrast, is diffuse, spread out, not to be found at a single point in space. This distinction seems perfectly clear. If you consider a cork bobbing up and down on an ocean swell, there is no doubt which is a particle and which a wave.

TABLE OV.V.1 Basic characteristics of particles and waves

Particles	Waves
Discrete	Continuous
Localized (here)	Nonlocalized (everywhere)
Individual	Collective

Yet as we enter the atomic realm, we will be faced with evidence that its inhabitants are not so easily classified. Electrons, protons, even light itself will be found to have characteristics of both particles *and* waves. In the strange world of quantum physics, "either-or" is replaced with "both-and." Rather than the wave-particle dichotomy of classical physics, we will come to recognize a *wave-particle duality* in quantum physics. The breakdown of the distinction between waves and particles undermines the Newtonian worldview, but at the same time it provides us with a richer and deeper understanding of nature.

20 Traveling Waves

This surfer is "catching a wave." At the same time, he is seeing light waves and hearing sound waves.

You may not realize it, but you are surrounded by waves. The "waviness" of a water wave is readily apparent, from the ripples on a pond to ocean waves large enough to surf. It's less apparent that sound and light are also waves because their wave properties are discovered only by careful observations and experiments. We will even find, quite surprisingly, that matter exhibits wave-like behavior when we get to the microscopic scale of electrons and atoms.

Our overarching goal in Part V is to understand the properties and characteristics that are common to waves of all types. In other words, we want to find the "essence of waviness" that all waves possess. In this chapter we start with the idea of a *traveling wave*. When your friend speaks to you, a sound wave travels through the air to your ear. Light waves travel from the sun to the earth. A sudden fracture in the earth's crust sends out a shock wave that is felt far away as an earthquake. To understand phenomena such as these we need both new models and new mathematics.

The boat's wake is a wave moving across the surface of the lake.

20.1 The Wave Model

The *particle model* of Parts I–IV focused on those aspects of motion that are common to many systems. Balls, cars, and rockets obviously differ from one another, but the general features of their motions are well described by treating them as particles. In Part V we will explore the basic properties of waves with a **wave model** that emphasizes those aspects of wave behavior common to all waves. Although water waves, sound waves, and light waves are clearly different, the wave model will allow us to understand many of their important features.

The wave model is built around the idea of a **traveling wave,** which is an organized disturbance that travels with a well-defined wave speed. This definition seems straightforward, but several new terms must be understood before the concept of a traveling wave will make sense.

To begin, we can distinguish three types of waves:

1. **Mechanical waves** can travel only within a material *medium,* such as air or water. Two mechanical waves that you are familiar with are sound waves and water waves.
2. **Electromagnetic waves,** which include visible light as well as radio waves, microwaves, and x rays, are a self-sustaining oscillation of the *electromagnetic field.* Electromagnetic waves require no material medium and can travel through a vacuum. We'll look more closely at electromagnetic waves in Section 20.5.
3. **Matter waves** are the basis for quantum physics. One of the most significant discoveries of the 20th century was that material particles, such as electrons and atoms, have wave-like characteristics. Chapter 24 will introduce matter waves.

The **medium** of a mechanical wave is the substance through or along which the wave moves. For example, the medium of a water wave is the water, the medium of a sound wave is the air, and the medium of a wave on a stretched string is the string. A medium must be *elastic.* That is, a restoring force of some sort brings the medium back to equilibrium after it has been displaced or disturbed. The tension in a stretched string pulls the string back straight after you pluck it. Gravity restores the level surface of a lake after the wave generated by a boat has passed by.

As a wave passes through a medium, the atoms that make up the medium— we'll simply call them the particles of the medium—are displaced from equilibrium. This is a **disturbance** of the medium. The water ripples of Figure 20.1 are a disturbance of the water's surface. A pulse traveling down a string is a disturbance, as is the wake of a boat and the sonic boom created by a jet traveling faster than the speed of sound. The disturbance of a wave is an *organized* motion of the particles in the medium, in contrast with the *random* molecular motions of thermal energy.

A wave disturbance is created by a *source.* The source of a wave might be a rock thrown into water, your hand plucking a stretched string, or an oscillating loudspeaker cone pushing on the air. Once created, the disturbance travels outward through the medium at the **wave speed** *v.* This is the speed with which a ripple moves across the water or a pulse travels down a string.

NOTE ▶ The disturbance propagates through the medium, and a wave does transfer *energy,* but **the medium as a whole does not move!** The ripples on the pond (the disturbance) move outward from the splash of the rock, but there is no outward flow of water from the splash. Likewise, the particles of a string oscillate up and down but do not move in the direction of a pulse traveling along the string. **A wave transfers energy, but it does not transfer any material or substance outward from the source.** ◀

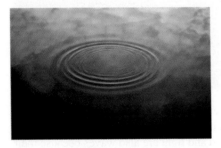

The disturbance is the rippling of the water's surface.

The water is the medium.

FIGURE 20.1 Ripples on a pond are a traveling wave.

We can identify two distinct types of wave motion: transverse and longitudinal.

Two types of wave motion

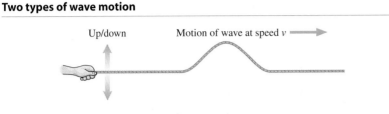

Up/down   Motion of wave at speed v →

A transverse wave

For mechanical waves, a **transverse wave** is a wave in which the particles in the medium move *perpendicular* to the direction in which the wave travels. For example, a wave travels along a string in a horizontal direction while the particles that make up the string oscillate vertically. Electromagnetic waves are also transverse waves because the electromagnetic fields oscillate perpendicular to the direction in which the wave travels.

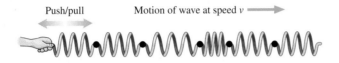

Push/pull   Motion of wave at speed v →

A longitudinal wave

In a **longitudinal wave,** the particles in the medium move *parallel* to the direction in which the wave travels. Here we see a chain of masses connected by springs. If you give the first mass in the chain a sharp push, a disturbance travels down the chain by compressing and expanding the springs. Sound waves in gases and liquids are the most well known examples of longitudinal waves. An oscillating loudspeaker cone compresses and expands the air much like the springs in this figure.

Some waves are more complex. For example, water waves have characteristics of both transverse and longitudinal waves. The surface of the water moves up and down vertically, but individual water molecules actually move both perpendicular *and* parallel to the direction of the wave. We will not analyze these more complex waves in this text.

Traveling Waves

How does a mechanical wave travel through a medium? In answering this question, we must be careful to distinguish the motion of the wave from the motion of the particles that make up the medium. The wave itself is not a particle, so we cannot apply Newton's laws to the wave. However, we can use Newton's laws to examine how the medium responds to a disturbance.

Figure 20.2 on the next page shows a transverse *wave pulse* traveling to the right along a stretched string. Imagine watching a little dot on the string as a wave pulse passes by. As the pulse approaches from the left, the string near the dot begins to curve. This is the **leading edge** of the pulse. Once the string curves, the tension forces pulling on a small segment of string no longer cancel each other. Instead, as you can see in the first part of the figure, the tension in the string exerts a net upward force on this segment of the string, causing it to accelerate upward.

The string's curvature reverses near the top of the pulse, causing the net force and the acceleration to point downward. Thus this piece of string slows until reaching a turning point at the highest point of the pulse, then it speeds up in the downward direction. This is much like a mass on a spring as the mass approaches the highest point and then reverses direction. Finally, the force and acceleration turn upward on the **trailing edge** of the pulse. The piece of string decelerates and comes to rest when the pulse has completely passed by.

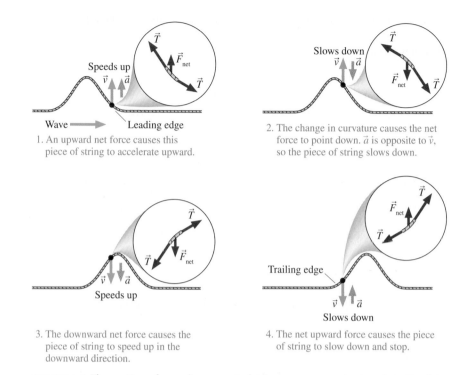

Speeds up

Wave → Leading edge

1. An upward net force causes this piece of string to accelerate upward.

Slows down

2. The change in curvature causes the net force to point down. $\vec{a}$ is opposite to $\vec{v}$, so the piece of string slows down.

Speeds up

3. The downward net force causes the piece of string to speed up in the downward direction.

Trailing edge

Slows down

4. The net upward force causes the piece of string to slow down and stop.

FIGURE 20.2 The motion of a small segment of string as a wave pulse travels to the right.

The point of this analysis is that no new physical principles are required to understand how a wave moves. The motion of a pulse along a string is a direct consequence of the tension acting on the segments of the string. An external force may have been required to create the pulse, but once started the pulse continues to move because of the *internal dynamics* of the medium.

This discovery leads to the important and somewhat surprising conclusion that the wave speed is a property *of the medium*. **The wave speed depends on the restoring forces within the medium but not at all on the shape or size of the pulse, how the pulse was generated, or how far it has traveled.** In Section 20.3 we'll apply the ideas of Figure 20.2 to derive an explicit expression for the wave speed on a string.

20.2 One-Dimensional Waves

Some waves travel in one dimension (waves on a string), some in two dimensions (ripples on a pond), and yet others in three dimensions (sound from a loudspeaker or light from the sun). We will begin our study with an analysis of one-dimensional waves, which are the easiest to visualize. The ideas developed in this section will carry over to two- and three-dimensional waves.

Waves on a String

The prototype of a one-dimensional wave is a wave on a string. Consider a string of total length L and total mass m. The ratio of mass to length is called the **linear density** μ of the string:

$$\mu = \frac{m}{L} \tag{20.1}$$

Linear density characterizes the *type* of string we are using. A fat string has a larger value of μ than a skinny string made of the same material. Similarly, a steel wire has a larger value of μ than a plastic string of the same diameter.

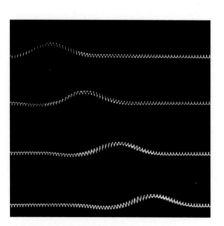

This sequence of photographs shows a wave pulse traveling along a string.

NOTE ▶ We'll assume that strings are *uniform,* meaning that the linear density is the same everywhere along the length of the string. ◀

As an example, the linear density of a 2.0-m-long string with a mass of 4.0 g is

$$\mu = \frac{0.0040 \text{ kg}}{2.0 \text{ m}} = 0.0020 \text{ kg/m}$$

The units tell us that the *numerical* value of μ is the mass in kg of a 1-m-long section of string; that is, the linear density is the mass *per meter* of string. But because μ is a ratio, we can apply it to a segment of string of any length. Thus the mass of any length L of string is $m = \mu L$.

The wave speed on a string depends on both the string's linear density μ and the tension T_s in the string. (The subscript s on the symbol T_s for the string's tension will distinguish it from the symbol T for the *period* of oscillation.) In Section 20.3 we will show that the wave speed on a stretched string is

Activ
ONLINE
Physics   10.2

$$v_{\text{string}} = \sqrt{\frac{T_s}{\mu}} \quad \text{(wave speed on a stretched string)} \quad (20.2)$$

This is the wave *speed,* not the wave velocity, so v_{string} always has a positive value.

Every point on a wave pulse travels with this speed. You can increase the wave speed either by *increasing* the string's tension (make it tighter) or by *decreasing* the string's linear density (make it skinnier). We'll examine the implications for stringed musical instruments in Chapter 21.

EXAMPLE 20.1 The speed of a wave pulse
A 2.0-m-long string with a mass of 4.0 g is tied to a wall at one end, stretched horizontally to a pulley 1.5 m away, then tied to a physics book of mass M that hangs from the string. Experiments find that a wave pulse travels along the stretched string at 40 m/s. What is the mass of the book?

MODEL The wave pulse is a traveling wave on a stretched string. The hanging book is in static equilibrium.

VISUALIZE Figure 20.3 shows the situation and a free-body diagram of the book.

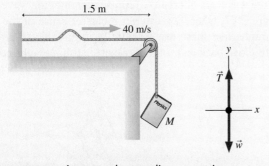

FIGURE 20.3 A wave pulse traveling on a string.

SOLVE The book is in static equilibrium, hence

$$(F_{\text{net}})_y = T_s - w = T_s - Mg = 0$$

Thus the tension in the string is $T_s = Mg$. The linear density of the 1.5-m-long segment of string between the wall and the pulley is exactly the same as μ for the entire string, which we computed in the calculation following Equation 20.1 to be 0.0020 kg/m. The length of the string between the wall and the pulley is not relevant. Squaring both sides of Equation 20.2 gives

$$v^2 = \frac{T_s}{\mu} = \frac{Mg}{\mu}$$

from which we find

$$M = \frac{\mu v^2}{g} = \frac{(0.0020 \text{ kg/m})(40 \text{ m/s})^2}{9.8 \text{ m/s}^2} = 0.327 \text{ kg} = 327 \text{ g}$$

ASSESS To be precise, 327 g is the combined mass of the book and the short length of the string that hangs from the pulley. This is one-quarter of the string, with a mass of 1 g, hence the book's mass is 326 g.

Snapshot Graphs and History Graphs

To understand waves we must deal with functions of *two* variables. Until now, we have been concerned with quantities that depend only on time, such as $x(t)$ or $v(t)$. Functions of the one variable t are all right for a particle, because a particle is only

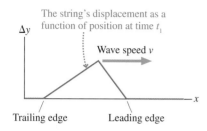

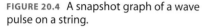

FIGURE 20.4 A snapshot graph of a wave pulse on a string.

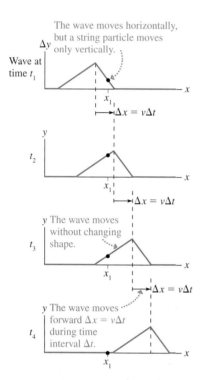

FIGURE 20.5 A sequence of snapshot graphs shows the wave in motion.

in one place at a time, but a wave is not localized. It is spread out through space at each instant of time. To describe a wave mathematically requires a function that specifies not only an instant of time (when) but also a point in space (where).

It will be helpful to think about waves graphically before we get into a mathematical analysis. Consider the wave pulse shown moving along a stretched string in Figure 20.4. (We will consider somewhat artificial triangular and square-shaped pulses in this section to make clear where the edges of the pulse are.) The graph shows the string's displacement Δy at a particular instant of time t_1 as a function of position x along the string. This is a "snapshot" of the wave, much like what you might make with a camera whose shutter is opened briefly at t_1. A graph that shows the wave's displacement as a function of position at a single instant of time is called a **snapshot graph.** For a wave on a string, a snapshot graph is literally a picture of the wave at this instant.

Figure 20.5 shows a sequence of snapshot graphs as the wave of Figure 20.4 continues to move. These are like successive frames from a movie. Notice that the wave pulse moves forward distance $\Delta x = v\Delta t$ during the time interval Δt. That is, the wave moves with constant speed.

A snapshot graph tells only half the story. It tells us *where* the wave is and how it varies with position, but only at one instant of time. It gives us no information about how the wave *changes* with time. As a different way of portraying the wave, suppose we follow the dot marked on the string in Figure 20.5 and produce a graph showing how the displacement of this dot changes with time. The result, shown below in Figure 20.6, is a displacement-versus-time graph at a single position in space. A graph that shows the wave's displacement as a function of time at a single position in space is called a **history graph.** It tells the history of that particular point in the medium.

You might think we have made a mistake; the graph of Figure 20.6 is reversed compared to Figure 20.5. It is not a mistake, but it requires careful thought to see why. As the wave moves toward the dot, the steep leading edge causes the dot to rise quickly. On the displacement-versus-time graph, *earlier* times (smaller values of t) are to the *left* and later times (larger t) to the right. Thus the leading edge of the wave is on the *left* side of the Figure 20.6 history graph. As you move to the right on Figure 20.6 you see the slowly falling trailing edge of the wave as it moves past the dot at later times.

The snapshot graph of Figure 20.4 and the history graph of Figure 20.6 portray complementary information. The snapshot graph tells us how things look throughout all of space, but at only one instant of time. The history graph tells us how things look at all times, but at only one position in space. We need them both to have the full story of the wave. An alternative representation of the wave is the series of graphs of Figure 20.7, where we can get a clearer sense of the wave moving forward. But graphs like these are essentially impossible to draw by hand, so it is necessary to move back and forth between snapshot graphs and history graphs.

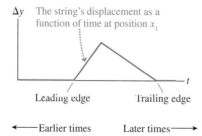

FIGURE 20.6 A history graph for the dot on the string in Figure 20.5.

FIGURE 20.7 An alternative look at a traveling wave.

EXAMPLE 20.2 Finding a history graph from a snapshot graph

Figure 20.8 is a snapshot graph at $t = 0$ s of a wave moving to the right at a speed of 2 m/s. Draw a history graph for the position $x = 8$ m.

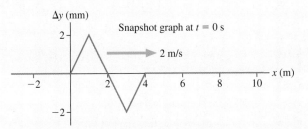

FIGURE 20.8 A snapshot graph at $t = 0$ s.

MODEL This is a wave traveling at constant speed. The pulse moves 2 m to the right every second.

VISUALIZE The snapshot graph of Figure 20.8 shows the wave at all points on the x-axis at $t = 0$ s. You can see that nothing is happening at $x = 8$ m at this instant of time because the wave has not yet reached $x = 8$ m. In fact, at $t = 0$ s the leading edge of the wave is still 4 m away from $x = 8$ m. Because the wave

is traveling at 2 m/s, it will take 2 s for the leading edge to reach $x = 8$ m. Thus the history graph for $x = 8$ m will be zero until $t = 2$ s. The first part of the wave causes a *downward* displacement of the medium, so immediately after $t = 2$ s the displacement at $x = 8$ m will be negative. The negative portion of the wave pulse is 2 m wide and takes 1 s to pass $x = 8$ m, so the midpoint of the pulse reaches $x = 8$ m at $t = 3$ s. The positive portion takes another 1 s to go past, so the trailing edge of the pulse arrives at $t = 4$ s. You could also note that the trailing edge was initially 8 m away from $x = 8$ m and needed 4 s to travel that distance at 2 m/s. The displacement at $x = 8$ m returns to zero at $t = 4$ s and remains zero for all later times. This information is all portrayed on the history graph of Figure 20.9.

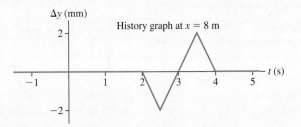

FIGURE 20.9 The corresponding history graph at $x = 8$ m.

STOP TO THINK 20.1 The graph at the right is the history graph at $x = 4$ m of a wave traveling to the right at a speed of 2 m/s. Which is the history graph of this wave at $x = 0$ m?

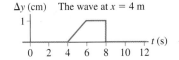

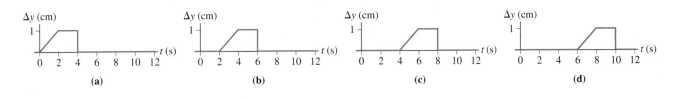

Longitudinal Waves

For a wave on a string, a transverse wave, the snapshot graph is literally a picture of the wave. Not so for a longitudinal wave, where the particles in the medium are displaced parallel to the direction in which the wave is traveling. Thus the displacement is Δx rather than Δy, and a snapshot graph is a graph of Δx versus x.

Figure 20.10a on the next page is a snapshot graph of a longitudinal wave, such as a sound wave. It's purposefully drawn to have the same shape as the string wave in Example 20.2. Without practice, what this graph tells us about the particles in the medium will not be obvious.

To help you find out, Figure 20.10b provides a tool for visualizing longitudinal waves. In the second row, we've used information from the graph to displace the particles in the medium to the right or to the left of their equilibrium positions. For example, the particle at $x = 1$ cm has been displaced 0.5 cm to the right because the snapshot graph shows $\Delta x = 0.5$ cm at $x = 1$ cm. We now have a picture of the longitudinal wave pulse at $t_1 = 0$ s. You can see that the medium is compressed to higher density at the center of the pulse and, to compensate, expanded

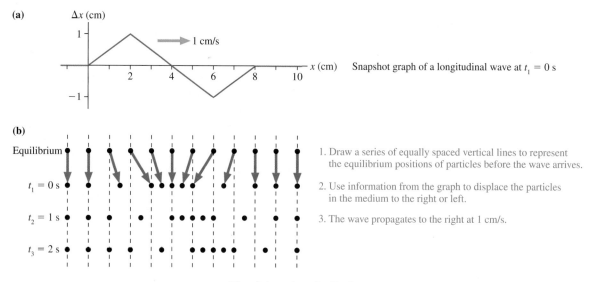

FIGURE 20.10 Visualizing a longitudinal wave.

to lower density at the leading and trailing edges. Two further lines show the medium at $t_2 = 1$ s and $t_3 = 2$ s so that you can see the wave propagating through the medium at 1 cm/s.

The Displacement

A traveling wave causes the particles of the medium to be displaced from their equilibrium positions. One of our goals is to develop a mathematical representation to describe all types of waves, so let's use the generic symbol D to stand for the *displacement* of a wave of any type. But what do we mean by a "particle" in the medium? And what about electromagnetic waves, for which there is no medium?

For a string, where the atoms stay fixed relative to each other, you can think of either the atoms themselves or very small segments of the string as being the particles of the medium. D is then the perpendicular displacement Δy of a point on the string. Sound waves and water waves require a bit more care because the atoms and molecules of a fluid have random thermal motions. Figure 20.10 is a nice way to visualize the motion of a sound wave, but no individual molecule in the air actually moves this way. For a fluid, we'll let a very small volume of the fluid be our particle. If the size of this little volume is large in comparison to the mean free path of the molecules, then the random thermal motions will average to zero. At the same time, the volume can be sufficiently small to be thought of as a "point" in the fluid.

Thus D for a sound wave is the longitudinal displacement Δx of a small volume of fluid. For any other mechanical wave, D is the appropriate displacement. Even electromagnetic waves can be described within the same mathematical representation if D is interpreted as a yet-undefined *electromagnetic field strength,* a "displacement" in a more abstract sense as an electromagnetic wave passes through a region of space.

> **NOTE** ▶ For convenience, we'll use the terminology of mechanical waves traveling through a medium. Even so, the mathematical representation of waves that we develop can be used to describe electromagnetic waves, and we will do so in Chapter 22. ◀

Because the displacement of a particle in the medium depends both on *where* the particle is (position x) and on *when* you observe it (time t), D must be a function of the two variables x and t. That is,

$$D(x, t) = \text{the displacement at time } t \text{ of a particle at position } x \qquad (20.3)$$

The values of *both* variables—where and when—must be specified before you can evaluate the displacement D.

Not every function of two variables describes a traveling wave. What we need is a function to describe a displacement that is translated along the x-axis at steady speed *without changing shape*. As a simple example, Figure 20.11a shows the parabola $f(x) = x^2$, which is centered at $x = 0$, and the *same parabola* drawn at $x = 2$ and at $x = 4$. The parabola at $x = 2$ can be written $f(x - 2) = (x - 2)^2$. Similarly, $f(x - 4) = (x - 4)^2$ is the parabola centered at $x = 4$.

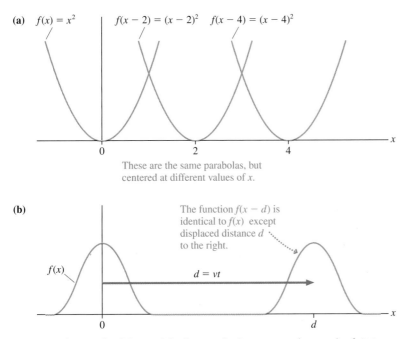

FIGURE 20.11 The graph of $f(x - d)$ looks exactly the same as the graph of $f(x)$ *except* that it is shifted distance d to the right.

In general, the graph of the function $f(x - d)$ looks exactly the same as the graph of $f(x)$ *except* that it is shifted distance d to the right. Whatever value $f(x)$ has at $x = 0$, the function $f(x - d)$ has the same value at $x = d$ where $x - d = 0$. Figure 20.11b applies this idea to a wave pulse. On the left is a snapshot graph of the wave at $t = 0$. At a later time t, the wave pulse looks *exactly the same* except that it has moved distance $d = vt$ to the right. If the wave at $t = 0$ is described by the function $f(x)$, then the shifted wave at time t is described by the function $f(x - d) = f(x - vt)$.

Thus we see that the displacement of a wave traveling in the positive x-direction with wave speed v must be a function of the form

$$D(x, t) = D(x - vt)$$

(wave traveling in the positive x-direction with speed v)   (20.4)

That is, the two variables x and t always appear together as $x - vt$. Examples of such functions include $D(x, t) = (x - vt)^2$, $D(x, t) = e^{(x - vt)}$, and $D(x, t) = \sin(x - vt)$. Each and every wave has its own function, depending on the shape and speed of the wave, so there is not a single function that describes all waves. But all waves traveling in the positive x-direction, regardless of their shape, are described by *some* function of the form $D(x - vt)$.

EXAMPLE 20.3 A traveling wave pulse

Draw snapshot graphs at $t = 0, 1,$ and 2 s to show that the displacement

$$D(x, t) = \frac{2 \text{ m}^3}{(x - (2 \text{ m/s})t)^2 + 1 \text{ m}^2}$$

where x is in m and t is in s, represents a traveling wave pulse.

VISUALIZE The snapshot graphs for $t = 0, 1,$ and 2 s are shown in Figure 20.12. You can see that the function $D(x, t)$ is a pulse that travels in the positive x-direction without changing shape. That is, it is a traveling wave with speed $v = 2$ m/s.

ASSESS The displacement is a function of $(x - (2 \text{ m/s})t)$. According to Equation 20.4, this should represent a wave traveling in the positive x-direction with speed $v = 2$ m/s. The graphs show that it does.

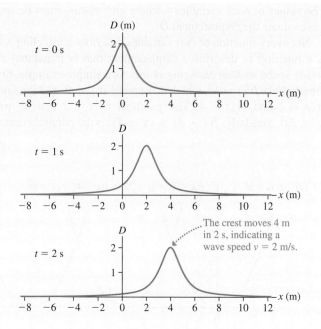

FIGURE 20.12 Three snapshot graphs of the displacement function of Example 20.3.

It should come as no surprise that the displacement of a wave traveling in the *negative x*-direction with wave speed v will have the form

$$D(x, t) = D(x + vt) \tag{20.5}$$

(wave traveling in the negative x-direction with speed v)

You can verify this with a homework problem.

> **NOTE** ▶ Not every wave moves along the x-axis. A change of variables will allow us to describe a wave traveling in the positive y-direction as $D(y - vt)$ and a wave traveling in the negative z-direction as $D(z + vt)$. We will usually let waves travel along the x-axis, just for convenience, but do not lose sight of the fact that we can equally well describe waves moving in any direction. ◀

STOP TO THINK 20.2   Which of the following actions would make a pulse travel faster down a stretched string? More than one answer may be correct. If so, give all that are correct.

a. Move your hand up and down more quickly as you generate the pulse.
b. Move your hand up and down a larger distance as you generate the pulse.
c. Use a heavier string of the same length, under the same tension.
d. Use a lighter string of the same length, under the same tension.
e. Stretch the string tighter to increase the tension.
f. Loosen the string to decrease the tension.
g. Put more force into the wave.

20.3 Sinusoidal Waves

10.1

A wave source that oscillates with simple harmonic motion (SHM) generates a **sinusoidal wave.** For example, a loudspeaker cone that oscillates in SHM radiates a sinusoidal sound wave. The sinusoidal electromagnetic waves broadcast by television and FM radio stations are generated by electrons oscillating back and forth in the antenna wire with SHM.

Figure 20.13 shows a sinusoidal wave moving through a medium. The source of the wave, which is undergoing vertical SHM, is located at $x = 0$. Notice how the wave crests move with steady speed toward larger values of x at later times t.

Figure 20.14a is a history graph for a sinusoidal wave, showing the displacement of the medium at one point in space. Each particle in the medium undergoes simple harmonic motion with frequency f, so this graph of SHM is identical to the graphs you learned to work with in Chapter 14. The *period* of the wave, shown on the graph, is the time interval for one cycle of the motion. The period is related to the wave *frequency f* by

$$T = \frac{1}{f} \tag{20.6}$$

exactly as in simple harmonic motion.

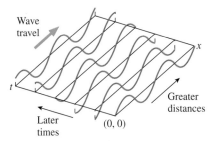

FIGURE 20.13 A sinusoidal wave moving along the x-axis.

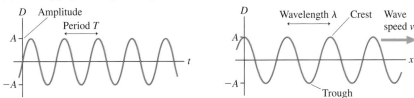

FIGURE 20.14 History and snapshot graphs for a sinusoidal wave.

Displacement-versus-time is only half the story. Figure 20.14b shows a snapshot graph for the same wave at one instant in time. Here we see the wave stretched out in space, moving to the right with speed v. The **amplitude** A of the wave is the maximum value of the displacement. The crests of the wave have displacement $D_{\text{crest}} = A$ and the troughs have displacement $D_{\text{trough}} = -A$.

An important characteristic of a sinusoidal wave is that it is periodic *in space* as well as in time. As you move from left to right along the "frozen" wave in the snapshot graph of Figure 20.14b, the disturbance repeats itself over and over. The distance spanned by one cycle of the motion is called the **wavelength** of the wave. Wavelength is symbolized by λ (lowercase Greek lambda) and, because it is a length, it is measured in units of meters. The wavelength is shown in Figure 20.14b as the distance between two crests, but it could equally well be the distance between two troughs.

> **NOTE ▶** Wavelength is the spatial analog of period. The period T is the *time* in which the disturbance at a single point in space repeats itself. The wavelength λ is the *distance* in which the disturbance at one instant of time repeats itself. ◀

The Fundamental Relationship for Sinusoidal Waves

There is an important relationship between the wavelength and the period of a wave. Figure 20.15 shows this relationship through five snapshot graphs of a sinusoidal wave at time increments of one-quarter of the period T. One full period has elapsed between the first graph and the last, which you can see by observing the motion at a fixed point on the x-axis. Each point in the medium has undergone exactly one complete oscillation.

The critical observation is that the wave crest marked by an arrow has moved one full wavelength between the first graph and the last. That is, **during a time interval of exactly one period T, each crest of a sinusoidal wave travels forward a distance of exactly one wavelength λ.** Because speed is distance divided by time, the wave speed must be

$$v = \frac{\text{distance}}{\text{time}} = \frac{\lambda}{T} \tag{20.7}$$

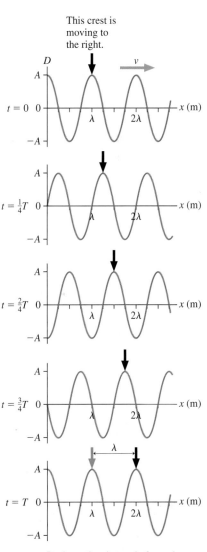

During a time interval of exactly one period, the crest has moved forward exactly one wavelength.

FIGURE 20.15 A series of snapshot graphs at time increments of one-quarter of the period T.

Because $f = 1/T$, it is customary to write Equation 20.7 in the form

$$v = \lambda f \qquad (20.8)$$

Although Equation 20.8 has no special name, it is *the* fundamental relationship for periodic waves. When using it, keep in mind the *physical* meaning that **a wave moves forward a distance of one wavelength during a time interval of one period.**

> **NOTE** ▶ Wavelength and period are defined only for *periodic* waves, so Equations 20.7 and 20.8 apply only to periodic waves. A wave pulse has a wave speed, but it doesn't have a wavelength or a period. Hence Equations 20.7 and 20.8 cannot be applied to wave pulses. ◀

STOP TO THINK 20.3 What is the frequency of this traveling wave?

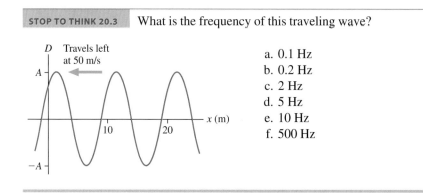

a. 0.1 Hz
b. 0.2 Hz
c. 2 Hz
d. 5 Hz
e. 10 Hz
f. 500 Hz

The Mathematics of Sinusoidal Waves

Section 20.2 introduced the function $D(x, t)$ that gives the displacement of a particle in the medium at position x and time t. We learned that the displacement must be a function of the form $D(x - vt)$ for a wave traveling in the positive x-direction with speed v. It's now straightforward to deduce the displacement function for a sinusoidal wave.

Figure 20.16 shows a snapshot graph at $t = 0$ of a sinusoidal wave. The sinusoidal function that describes the displacement of this wave is

$$D(x, t = 0) = A \sin\left(2\pi \frac{x}{\lambda} + \phi_0\right) \qquad (20.9)$$

where the notation $D(x, t = 0)$ means that we've frozen the time at $t = 0$ to make the displacement a function only of x. The term ϕ_0 is a *phase constant* that characterizes the initial conditions. (We'll return to the phase constant momentarily.)

The function of Equation 20.9 is periodic with period λ. We can see this by writing

$$D(x + \lambda) = A \sin\left(2\pi \frac{(x + \lambda)}{\lambda} + \phi_0\right) = A \sin\left(2\pi \frac{x}{\lambda} + \phi_0 + 2\pi \text{ rad}\right)$$

$$= A \sin\left(2\pi \frac{x}{\lambda} + \phi_0\right) = D(x)$$

where we used the fact that $\sin(a + 2\pi \text{ rad}) = \sin a$. In other words, the disturbance created by the wave at $x + \lambda$ is exactly the same as the disturbance at x.

Now, just as we did in the previous section for Figure 20.11b, we can set the wave in motion by replacing x in Equation 20.9 with $x - vt$. This will cause the

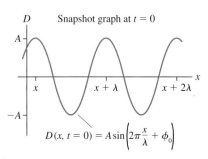

FIGURE 20.16 A sinusoidal wave is "frozen" at $t = 0$.

wave to be displaced distance $d = vt$ to the right at time t but with no change in shape. Thus the displacement equation of a sinusoidal wave is

$$D(x, t) = A\sin\left(2\pi\frac{x - vt}{\lambda} + \phi_0\right) = A\sin\left(2\pi\left(\frac{x}{\lambda} - \frac{t}{T}\right) + \phi_0\right) \quad (20.10)$$

In the last step we used $v = \lambda f = \lambda/T$ to write $v/\lambda = 1/T$. The function of Equation 20.10 is not only periodic in space with period λ, it is also periodic in time with period T. That is, $D(x, t + T) = D(x, t)$.

It will be useful to introduce two new quantities. First, recall from simple harmonic motion the *angular frequency*

$$\omega = 2\pi f = \frac{2\pi}{T} \quad (20.11)$$

The units of ω are rad/s, although many textbooks use simply s^{-1}.

You can see that ω is 2π times the reciprocal of the period in time. This suggests that we define an analogous quantity, called the **wave number** k, that is 2π times the reciprocal of the period in space:

$$k = \frac{2\pi}{\lambda} \quad (20.12)$$

The units of k are rad/m, although many textbooks use simply m^{-1}.

NOTE ▶ The wave number k is *not* a spring constant, even though it uses the same symbol. This is a most unfortunate use of symbols, but every major textbook and professional tradition uses the same symbol k for these two very different meanings, so we have little choice but to follow along. ◀

We can use the fundamental relationship $v = \lambda f$ to find an analogous relationship between ω and k:

$$v = \lambda f = \frac{2\pi}{k}\frac{\omega}{2\pi} = \frac{\omega}{k}$$

which is usually written

$$\omega = vk \quad (20.13)$$

Equation 20.13 contains no new information. It is a variation of Equation 20.8, but one that is convenient when working with k and ω.

If we use the definitions of Equations 20.11 and 20.12, Equation 20.10 for the displacement can be written

$$D(x, t) = A\sin(kx - \omega t + \phi_0)$$
(sinusoidal wave traveling in the positive x-direction)
$\quad (20.14)$

A sinusoidal wave traveling in the negative x-direction, would be $A\sin(kx + \omega t + \phi_0)$. Equation 20.14 is graphed versus x and t in Figure 20.17.

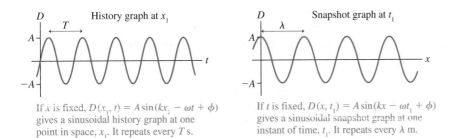

If x is fixed, $D(x_1, t) = A\sin(kx_1 - \omega t + \phi)$ gives a sinusoidal history graph at one point in space, x_1. It repeats every T s.

If t is fixed, $D(x, t_1) = A\sin(kx - \omega t_1 + \phi)$ gives a sinusoidal snapshot graph at one instant of time, t_1. It repeats every λ m.

FIGURE 20.17 Interpreting the equation of a sinusoidal traveling wave.

Just as it did for simple harmonic motion, the phase constant ϕ_0 characterizes the initial conditions. At $(x, t) = (0\text{ m}, 0\text{ s})$ Equation 20.14 becomes

$$D(0\text{ m}, 0\text{ s}) = A\sin\phi_0 \qquad (20.15)$$

Different values of ϕ_0 describe different initial conditions for the wave.

EXAMPLE 20.4 Analyzing a sinusoidal wave

A sinusoidal wave with an amplitude of 1.0 cm and a frequency of 100 Hz travels at 200 m/s in the positive x-direction. At $t = 0$ s, the point $x = 1.0$ m is on a crest of the wave.

a. Determine the values of A, v, λ, k, f, ω, T, and ϕ_0 for this wave.
b. Write the equation for the wave's displacement as it travels.
c. Draw a snapshot graph of the wave at $t = 0$ s.

VISUALIZE The snapshot graph will be sinusoidal, but we must do some numerical analysis before we know how to draw it.

SOLVE

a. There are several numerical values associated with a sinusoidal traveling wave, but they are not all independent. From the problem statement itself we learn that

$$A = 1.0\text{ cm} \qquad v = 200\text{ m/s} \qquad f = 100\text{ Hz}$$

We can then find:

$$\lambda = v/f = 2.00\text{ m}$$

$$k = 2\pi/\lambda = \pi\text{ rad/m} \quad\text{or}\quad 3.14\text{ rad/m}$$

$$\omega = 2\pi f = 628\text{ rad/s}$$

$$T = 1/f = 0.0100\text{ s} = 10.0\text{ ms}$$

The phase constant ϕ_0 is determined by the initial conditions. We know that a wave crest, with displacement $D = A$, is passing $x_0 = 1.0$ m at $t_0 = 0$ s. Equation 20.14 at x_0 and t_0 is

$$D(x_0, t_0) = A = A\sin(k(1.0\text{ m}) + \phi_0)$$

This equation is true only if $\sin(k(1.0\text{ m}) + \phi_0) = 1$, which requires

$$k(1.0\text{ m}) + \phi_0 = \frac{\pi}{2}\text{rad}$$

Solving for the phase constant gives

$$\phi_0 = \frac{\pi}{2}\text{rad} - (\pi\text{rad/m})(1.0\text{ m}) = -\frac{\pi}{2}\text{rad}$$

b. With the information gleaned from part a, the wave's displacement is

$$D(x, t) = 1.0\text{ cm} \times$$

$$\sin\left[(3.14\text{ rad/m})x - (628\text{ rad/s})t - \frac{\pi}{2}\text{ rad}\right]$$

Notice that we included units with A, k, ω, and ϕ_0.

c. We know that $x = 1.0$ m is a wave crest at $t = 0$ s and that the wavelength is $\lambda = 2.0$ m. Because the origin is $\lambda/2$ away from the crest at $x = 1.0$ m, we expect to find a wave trough at $x = 0$. This is confirmed by calculating $D(0\text{ m}, 0\text{ s}) = (1.0\text{ cm})\sin(-\pi/2\text{ rad}) = -1.0\text{ cm}$. Figure 20.18 is a snapshot graph that portrays this information.

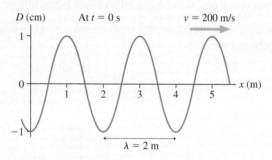

FIGURE 20.18 A snapshot graph at $t = 0$ s of the sinusoidal wave of Example 20.4.

Wave Motion on a String

The displacement equation, Equation 20.14, allows us to learn more about wave motion on a string. As a wave travels along the x-axis, the points on the string oscillate back and forth in the y-direction. The displacement D of a point on the string is simply that point's y-coordinate, so Equation 20.14 for a string wave is

$$y(x, t) = A\sin(kx - \omega t + \phi_0) \qquad (20.16)$$

The velocity of a particle on the string—**which is not the same as the velocity of the wave along the string**—is the time derivative of Equation 20.16:

$$v_y = \frac{dy}{dt} = -\omega A\cos(kx - \omega t + \phi_0) \qquad (20.17)$$

The maximum velocity of a small segment of the string is $v_{\max} = \omega A$. This is the same result we found for simple harmonic motion because the motion of the string particles *is* simple harmonic motion. Figure 20.19 shows velocity vectors *of the particles* at different points on a sinusoidal wave.

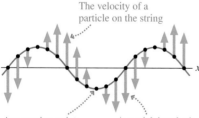

The velocity of the wave ⟶

The velocity of a particle on the string

At a turning point, the particle has zero velocity.

A particle's velocity is maximum at zero displacement.

FIGURE 20.19 A snapshot graph of a wave on a string with vectors showing the velocity *of the string* at various points.

NOTE ▶ Creating a wave of larger amplitude increases the speed of particles *in* the medium, but it does *not* change the speed of the wave *through* the medium. ◀

Pursuing this line of thought, we can derive an expression for the wave speed along the string. Figure 20.20 shows a small segment of the string with length $\Delta x \ll \lambda$ right at a crest of the wave. You can see that the string's tension exerts a downward force on this piece of the string, pulling it back to equilibrium. Newton's second law for this small segment of string is

$$(F_{\text{net}})_y = ma_y = (\mu \Delta x)a_y \tag{20.18}$$

where we used the string's linear density μ to write the mass as $m = \mu \Delta x$.

From simple harmonic motion, we know that this point of maximum displacement is also the point of maximum acceleration. The acceleration of a point on the string is the time derivative of Equation 20.17:

$$a_y = \frac{dv_y}{dt} = -\omega^2 A \sin(kx - \omega t + \phi_0) \tag{20.19}$$

Thus the acceleration at the crest of the wave is $a_y = -\omega^2 A$. But the angular frequency ω with which the particles of the string oscillate is related to the wave's speed v along the string by Equation 20.13, $\omega = vk$. Thus

$$a_y = -\omega^2 A = -v^2 k^2 A \tag{20.20}$$

A large wave speed causes the particles of the string to oscillate more quickly and thus to have a larger acceleration.

Our main task is to determine the net force acting on this segment of the string. You can see from Figure 20.20 that the y-component of the tension is $T_s \sin \theta$, where θ is the angle of the string at $x = \frac{1}{2}\Delta x$. θ is a *negative* angle because it is below the x-axis. This segment of string is pulled from both ends, so

$$(F_{\text{net}})_y = 2T_s \sin \theta \tag{20.21}$$

The angle θ is a very small angle, because $\Delta x \ll \lambda$, so we can use the small-angle approximation ($\sin u \approx \tan u$ if $u \ll 1$) to write

$$(F_{\text{net}})_y \approx 2T_s \tan \theta \tag{20.22}$$

where $\tan \theta$ is the slope of the string at $x = \frac{1}{2}\Delta x$.

At this specific instant, with the crest of the wave at $x = 0$, the equation of the string is

$$y = A \cos(kx)$$

The slope of the string at $x = \frac{1}{2}\Delta x$ is the derivative evaluated at that point:

$$\tan \theta = \frac{dy}{dx}\bigg|_{\text{at } \Delta x/2} = -kA \sin(kx)|_{\text{at } \Delta x/2} = -kA \sin\left(\frac{k\Delta x}{2}\right)$$

Now $\Delta x \ll \lambda$, so $k\Delta x/2 = \pi \Delta x/\lambda \ll 1$. Thus the small-angle approximation ($\sin u \approx u$ if $u \ll 1$) of the slope is

$$\tan \theta \approx -kA\left(\frac{k\Delta x}{2}\right) = -\frac{k^2 A \Delta x}{2} \tag{20.23}$$

If we substitute this expression for $\tan \theta$ into Equation 20.22, we find that the net force on this little piece of string is

$$(F_{\text{net}})_y = -k^2 A T_s \Delta x \tag{20.24}$$

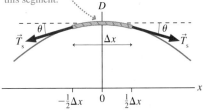

A small segment of the string at the crest of the wave. Because of the curvature of the string, the tension forces exert a net downward force on this segment.

FIGURE 20.20 A small segment of string at the crest of a wave.

Now we can use Equation 20.20 for a_y and Equation 20.24 for $(F_{net})_y$ in Newton's second law. With these substitutions, Equation 20.18 becomes

$$(F_{net})_y = -k^2 A T_s \Delta x = (\mu \Delta x) a_y = -v^2 k^2 A \mu \Delta x \qquad (20.25)$$

The term $-k^2 A \Delta x$ cancels, and we're left with

$$v = \sqrt{\frac{T_s}{\mu}} \qquad (20.26)$$

This was the result that we stated, without proof, in Equation 20.2. Although we've derived Equation 20.26 with the assumption of a sinusoidal wave, the wave speed does not depend on the shape of the wave. Thus any wave on a stretched string will have this wave speed.

EXAMPLE 20.5 Generating a sinusoidal wave

A very long string with $\mu = 2.0$ g/m is stretched along the x-axis with a tension of 5.0 N. At $x = 0$ m it is tied to a 100 Hz simple harmonic oscillator that vibrates perpendicular to the string with an amplitude of 2.0 mm. The oscillator is at its maximum positive displacement at $t = 0$ s.

a. Write the displacement equation for the traveling wave on the string.

b. At $t = 5.0$ ms, what is the string's displacement at a point 2.7 m from the oscillator?

MODEL The oscillator generates a sinusoidal traveling wave on a string. The displacement of the wave has to match the displacement of the oscillator at $x = 0$ m.

SOLVE

a. The equation for the displacement is

$$D(x, t) = A \sin(kx - \omega t + \phi_0)$$

with A, k, ω, and ϕ_0 to be determined. The wave amplitude is the same as the amplitude of the oscillator that generates the wave, so $A = 2.0$ mm. The oscillator has its maximum displacement $y_{osc} = A = 2.0$ mm at $t = 0$ s, thus

$$D(0 \text{ m}, 0 \text{ s}) = A \sin(\phi_0) = A$$

This requires the phase constant to be $\phi_0 = \pi/2$ rad. The wave's frequency is $f = 100$ Hz, the frequency of the source, therefore the angular frequency is $\omega = 2\pi f = 200\pi$ rad/s.

We still need $k - 2\pi/\lambda$, but we do not know the wavelength. However, we have enough information to determine the wave speed, and we can then use either $\lambda = v/f$ or $k = \omega/v$. The speed is

$$v = \sqrt{\frac{T_s}{\mu}} = \sqrt{\frac{5.0 \text{ N}}{0.0020 \text{ kg/m}}} = 50 \text{ m/s}$$

Using v, we find $\lambda = 0.50$ m and $k = 2\pi/\lambda = 4\pi$ rad/m. Thus the wave's displacement is

$$D(x, t) = (2.0 \text{ mm}) \times$$

$$\sin\left[2\pi((2.0 \text{ m}^{-1})x - (100 \text{ s}^{-1})t) + \frac{\pi}{2} \text{ rad}\right]$$

where x is in m and t in s. Notice that we have separated out the 2π. This step is not essential, but for some problems it makes subsequent steps easier.

b. The wave's displacement at $t = 5.0$ ms $= 0.0050$ s is

$$D(x, t = 5.0 \text{ ms}) = (2.0 \text{ mm}) \sin\left(4\pi x - \pi + \frac{\pi}{2} \text{ rad}\right)$$

$$= (2.0 \text{ mm}) \sin\left(4\pi x - \frac{\pi}{2} \text{ rad}\right)$$

At $x = 2.7$ m (calculator set to radians!), the displacement is

$$D(2.7 \text{ m}, 5.0 \text{ ms}) = 1.62 \text{ mm}$$

20.4 Waves in Two and Three Dimensions

Suppose you were to take a photograph of ripples spreading on a pond. If you mark the location of the *crests* on the photo, your picture would look like Figure 20.21a. The lines that locate the crests are called **wave fronts,** and they are spaced precisely one wavelength apart. The diagram shows only a single instant of time, but you can imagine a movie in which you would see the wave fronts moving outward from the source at speed v. A wave like this is called a **circular wave.** It is a two-dimensional wave that spreads across a surface.

Although the wave fronts are circles, you would hardly notice the curvature if you observed a small section of the wave front very, very far away from the source. The wave fronts would appear to be parallel lines, still spaced one wavelength apart and traveling at speed v. A good example is an ocean wave reaching a beach. Ocean waves are generated by storms and wind far out at sea, hundreds or thousands of miles away. By the time they reach the beach where you are working on your tan, the crests appear to be straight lines. An aerial view of the ocean would show a wave diagram like Figure 20.21b.

Many waves of interest, such as sound waves or light waves, move in three dimensions. For example, loudspeakers and light bulbs emit **spherical waves.** That is, the crests of the wave form a series of concentric spherical shells separated by the wavelength λ. In essence, the waves are three-dimensional ripples. It will still be useful to draw wave-front diagrams such as Figure 20.21, but now the circles are slices through the spherical shells locating the wave crests.

If you observe a spherical wave very, very far from its source, the small piece of the wave front that you can see is a little patch on the surface of a very large sphere. If the radius of the sphere is sufficiently large, you will not notice the curvature and this little patch of the wave front appears to be a plane. Figure 20.22 illustrates the idea of a **plane wave.**

To visualize a plane wave, imagine standing on the x-axis facing a sound wave as it comes toward you from a very distant loudspeaker. Sound is a longitudinal wave, so the particles of medium oscillate toward you and away from you. If you were to locate all of the particles that, at one instant of time, were at their maximum displacement toward you, they would all be located in a plane perpendicular to the travel direction. This is one of the wave fronts in Figure 20.22, and all the particles in this plane are doing exactly the same thing at that instant of time. This plane is moving toward you at speed v. There is another plane one wavelength behind it where the molecules are also at maximum displacement, yet another two wavelengths behind the first, and so on.

Because a plane wave's displacement depends on x but not on y or z, the displacement function $D(x, t)$ describes a plane wave just as readily as it does a one-dimensional wave. Once you specify a value for x, the displacement is the same at every point in the yz-plane that slices the x-axis at that value (i.e., one of the planes shown in Figure 20.22).

NOTE ▶ There are no perfect plane waves in nature, but it is a good model for many waves of practical interest. ◀

We can describe a circular wave or a spherical wave by changing the mathematical description from $D(x, t)$ to $D(r, t)$, where r is the distance measured outward from the source. Then the displacement of the medium will be the same at every point on a spherical surface. In particular, a sinusoidal spherical wave with wave number k and angular frequency ω is written

$$D(r, t) = A(r) \sin(kr - \omega t + \phi_0) \qquad (20.27)$$

Other than the change of x to r, the only difference is that the amplitude is now a function of r. A one-dimensional wave propagates with no change in the wave amplitude. But circular and spherical waves spread out to fill larger and larger volumes of space. To conserve energy, an issue we'll look at later in the chapter, the wave's amplitude has to decrease with increasing distance r. This is why sound and light decrease in intensity as you get farther from the source. We don't need to specify exactly how the amplitude decreases with distance, but you should be aware that it does.

Phase and Phase Difference

The quantity $(kx - \omega t + \phi_0)$ is called the **phase** of the wave, denoted ϕ. The phase of a wave will be an important concept in Chapters 21 and 22, where we will explore the consequences of adding various waves together. For now, we can

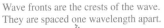

(a)

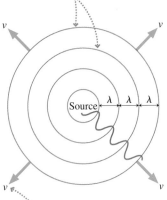

Wave fronts are the crests of the wave. They are spaced one wavelength apart.

The circular wave fronts move outward from the source at speed v.

(b)

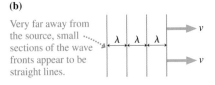

Very far away from the source, small sections of the wave fronts appear to be straight lines.

FIGURE 20.21 The wave fronts of a circular or spherical wave.

Very far from the source, small segments of spherical wave fronts appear to be planes. The wave is cresting at every point in these planes.

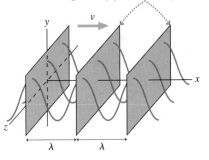

FIGURE 20.22 A plane wave.

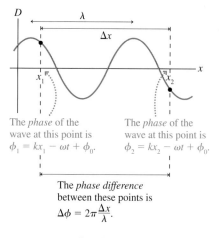

The *phase* of the wave at this point is $\phi_1 = kx_1 - \omega t + \phi_0$.

The *phase* of the wave at this point is $\phi_2 = kx_2 - \omega t + \phi_0$.

The *phase difference* between these points is $\Delta\phi = 2\pi\dfrac{\Delta x}{\lambda}$.

FIGURE 20.23 The phase difference between two points on a wave.

note that the wave fronts seen in Figures 20.21 and 20.22 are "surfaces of constant phase." To see this, notice that we can use the phase to write the displacement as simply $D(x, t) = A \sin\phi$. Because each point on a wave front has the same displacement, the phase must be the same at every point.

It will be useful to know the *phase difference* $\Delta\phi$ between two different points on a sinusoidal wave. Figure 20.23 shows two points on a sinusoidal wave at time t. The phase difference between these point is

$$\Delta\phi = \phi_2 - \phi_1 = (kx_2 - \omega t + \phi_0) - (kx_1 - \omega t + \phi_0)$$

$$= k(x_2 - x_1) = k\Delta x = 2\pi\frac{\Delta x}{\lambda} \qquad (20.28)$$

That is, **the phase difference between two points on a wave depends on the ratio of their separation Δx to the wavelength λ**. For example, two points on a wave separated by $\Delta x = \frac{1}{2}\lambda$ have a phase difference $\Delta\phi = \pi$.

An important consequence of Equation 20.28 is that **the phase difference between two adjacent wave fronts is $\Delta\phi = 2\pi$**. This follows from the fact that two adjacent wave fronts are separated by $\Delta x = \lambda$. This is an important idea. Moving from one crest of the wave to the next corresponds to changing the *distance* by λ and to changing the *phase* by 2π.

EXAMPLE 20.6 The phase difference between two points on a sound wave
A 100 Hz sound wave travels with a wave speed of 343 m/s.

a. What is the phase difference between two points 60 cm apart along the direction the wave is traveling?
b. How far apart are two points whose phase differs by 90°?

MODEL Treat the wave as a plane wave traveling in the positive x-direction.

SOLVE

a. The phase difference between two points at positions x_1 and x_2 is

$$\Delta\phi = 2\pi\frac{\Delta x}{\lambda}$$

In this case, $\Delta x = 60$ cm $= 0.60$ m. The wavelength is

$$\lambda = \frac{v}{f} = \frac{343 \text{ m/s}}{100 \text{ Hz}} = 3.43 \text{ m}$$

and thus

$$\Delta\phi = 2\pi\frac{0.60 \text{ m}}{3.43 \text{ m}} = 0.350\pi \text{ rad} = 63.0°$$

b. A phase difference $\Delta\phi = 90°$ is $\pi/2$ radians. This will be the phase difference between two points when $\Delta x/\lambda = \frac{1}{4}$, or when $\Delta x = \lambda/4$. In this case, with $\lambda = 3.43$ m, $\Delta x = 85.8$ cm.

ASSESS The phase difference increases as Δx increases, so we expect the answer to part b to be larger than 60 cm.

STOP TO THINK 20.4 What is the phase difference between the crest of a wave and the adjacent trough?

a. -2π b. 0
c. $\pi/4$ d. $\pi/2$
e. π f. 3π

20.5 Sound and Light

Although there are many kinds of waves in nature, two are especially significant for us as humans. These are sound waves and light waves, the basis of hearing and seeing.

Sound Waves

We usually think of sound waves traveling in air, but sound can travel through any gas, through liquids, and even through solids. Figure 20.24 shows a loudspeaker cone vibrating back and forth in a fluid such as air or water. Each time the cone moves forward, it collides with the molecules and pushes them closer together. A half cycle later, as the cone moves backward, the fluid has room to expand and the density decreases a little. These regions of higher and lower density are called **compressions** and **rarefactions.**

This periodic sequence of compressions and rarefactions travels outward from the loudspeaker as a longitudinal sound wave. A similar type of sound wave is produced if you hit the end of a metal rod with a hammer, sending a compression pulse through the metal.

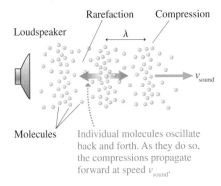

> **NOTE** ▶ Sound waves in gases and liquids are always longitudinal waves, but sound waves in solids can be either longitudinal or transverse. For a transverse wave to propagate, a plane of molecules oscillating perpendicular to the direction of motion has to be able to "drag" the neighboring planes of atoms along with it. Neighboring planes slip in a gas or liquid, so these media won't support a transverse wave. (Think how much easier it is to slide your hand sideways in water than to push against the water.) But the stronger molecular bonds in a solid do support transverse sound waves, sometimes called *shear waves*. Their speed is usually different from the speed of longitudinal sound waves. We'll assume that all sound waves are longitudinal waves unless otherwise noted. ◀

The speed of sound waves depends on the properties of the medium. A thermodynamic analysis of the compressions and expansions shows that the wave speed in a gas depends on the temperature and on the molecular mass of the gas. For air at room temperature (20°C),

$$v_{sound} = 343 \text{ m/s} \qquad \text{(sound speed in air at 20°C)}$$

The speed of sound is a little bit lower at lower temperatures and a little higher at higher temperatures. Liquids and solids are less compressible than air, and that makes the speed of sound in those media higher than in air. Table 20.1 gives the speed of sound in several substances.

A speed of 343 m/s is high, but not extraordinarily so. A distance as small as 100 m is enough to notice a slight delay between when you see something, such as a person hammering a nail, and when you hear it. The time required for sound to travel 1 km is $t = (1000 \text{ m})/(343 \text{ m/s}) \approx 3$ s. You may have learned to estimate the distance to a bolt of lightning by timing the number of seconds between when you see the flash and when you hear the thunder. Because sound takes 3 s to travel 1 km, the time divided by 3 gives the distance in kilometers. Or, in English units, the time divided by 5 gives the distance in miles.

Your ears are able to detect sinusoidal sound waves with frequencies between about 20 Hz and about 20,000 Hz, or 20 kHz. Low frequencies are perceived as a "low pitch" bass note while high frequencies are heard as a "high pitch" treble note. Your high-frequency range of hearing can deteriorate either with age or as a result of exposure to loud sounds that damage the ear.

Sound waves exist at frequencies well above 20 kHz, even though humans can't hear them. These are called *ultrasonic* frequencies. Oscillators vibrating at frequencies of many MHz generate the ultrasonic waves used in ultrasound medical imaging. A 3 MHz frequency traveling through water (which is basically what your body is) at a sound speed of 1480 m/s has a wavelength of about 0.5 mm. It is this very small wavelength that allows ultrasound to image very small objects. We'll see why when we study *diffraction* in Chapter 22.

FIGURE 20.24 A sound wave in a fluid is a sequence of compressions and rarefactions that travels outward with speed v_{sound}. The variation in density and the amount of motion have been greatly exaggerated.

10.3

TABLE 20.1 The speed of sound

Medium	Speed (m/s)
Air (0°C)	331
Air (20°C)	343
Helium (0°C)	970
Ethyl alcohol	1170
Water	1480
Granite	6000
Aluminum	6420

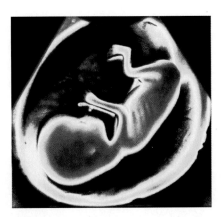

This ultrasound image is an example of how high-frequency sound waves can be used to "see" within the human body.

EXAMPLE 20.7 Sound wavelengths

What are the wavelengths of sound waves at the limits of human hearing and at the midrange frequency of 500 Hz? Notes sung by human voices are near 500 Hz, as are notes played by striking keys near the center of a piano keyboard.

MODEL Assume a room temperature 20°C.

SOLVE We can use the fundamental relationship $\lambda = v/f$ to find the wavelengths for sounds of various frequencies:

$$f = 20 \text{ Hz} \qquad \lambda = \frac{343 \text{ m/s}}{20 \text{ Hz}} = 17 \text{ m}$$

$$f = 500 \text{ Hz} \qquad \lambda = \frac{343 \text{ m/s}}{500 \text{ Hz}} = 0.69 \text{ m}$$

$$f = 20,000 \text{ Hz} \qquad \lambda = \frac{343 \text{ m/s}}{20,000 \text{ Hz}} = 0.017 \text{ m} = 1.7 \text{ cm}$$

ASSESS The wavelength of a 20 kHz note is a small 1.7 cm while, at the other extreme, a 20 Hz note has a huge wavelength of 17 m! This is because a wave moves forward one wavelength during a time interval of one period, and a wave traveling at 343 m/s can move 17 m during the $\frac{1}{20}$ s period of a 20 Hz note. The 69 cm wavelength of a 500 Hz note is more of a "human scale." You might note that most musical instruments are a meter or a little less in size. This is not a coincidence. You will see in the next chapter how the wavelength produced by a musical instrument is related to its size.

Electromagnetic Waves

A light wave is an *electromagnetic wave,* an oscillation of the electromagnetic field. Other electromagnetic waves, such as radio waves, microwaves, and ultraviolet light, have the same physical characteristics as light waves even though we cannot sense them with our eyes. It is easy to demonstrate that light will pass unaffected through a container from which all the air has been removed, and light reaches us from distant stars through the vacuum of interstellar space. Such observations raise interesting but difficult questions. If light can travel through a region in which there is no matter, then what is the *medium* of a light wave? What is it that is waving?

It took scientists over 50 years, most of the 19th century, to answer this question. We will examine the answers in more detail in Part VI after we introduce the ideas of electric and magnetic fields. For now we can say that light waves are a "self-sustaining oscillation of the electromagnetic field." Being self-sustaining means that electromagnetic waves require *no material medium* in order to travel, hence electromagnetic waves are not mechanical waves. Fortunately, we can learn about the wave properties of light without having to understand electromagnetic fields. In fact, the discovery that light propagates as a wave was made 60 years before it was realized that light is an electromagnetic wave. We, too, will be able to learn much about the wave nature of light without having to know just what it is that is waving.

It was predicted theoretically in the late 19th century, and has been subsequently confirmed experimentally with outstanding precision, that all electromagnetic waves travel through vacuum with the same speed, called the *speed of light.* The value of the speed of light is

$$v_{\text{light}} = c = 299{,}792{,}458 \text{ m/s} \qquad \text{(electromagnetic wave speed in vacuum)}$$

where the special symbol c is used to designate the speed of light. (This is the c in Einstein's famous formula $E = mc^2$.) Now *this* is really moving—about one million times faster than the speed of sound in air! At this speed, light could circle the earth 7.5 times in a mere one second—*if* there were some way to make it go in circles.

NOTE ▶ $c = 3.00 \times 10^8$ m/s is the appropriate value to use in most calculations. ◀

The wavelengths of light are extremely small. You will learn in Chapter 22 how these wavelengths are determined, but for now we will note that visible light is an electromagnetic wave with a wavelength (in air) in the range of roughly 400 nm (400×10^{-9} m) to 700 nm (700×10^{-9} m). Each wavelength is perceived as a different color, with the longer wavelengths seen as orange or red light and the shorter wavelengths seen as blue or violet light. A prism is able to spread the different wavelengths apart, from which we learn that "white light" is all the colors, or wavelengths, combined. The spread of colors seen with a prism, or seen in a rainbow, is called the *visible spectrum.*

If the wavelengths of light are unbelievably small, the oscillation frequencies are unbelievably large. The frequency for a 600 nm wavelength of light is

$$f = \frac{v}{\lambda} = \frac{3.00 \times 10^8 \text{ m/s}}{600 \times 10^{-9} \text{ m}} = 5.00 \times 10^{14} \text{ Hz}$$

The frequencies of light waves are roughly a factor of a trillion (10^{12}) higher than sound frequencies.

Electromagnetic waves exist at many frequencies other than the rather limited range that our eyes detect. One of the major technological advances of the twentieth century was learning to generate and detect electromagnetic waves at many frequencies, ranging from low-frequency radio waves to the extraordinarily high frequencies of x rays. Figure 20.25 shows that the visible spectrum is a small slice out of the much broader **electromagnetic spectrum.**

White light passing through a prism is spread out into a band of colors called the *visible spectrum.*

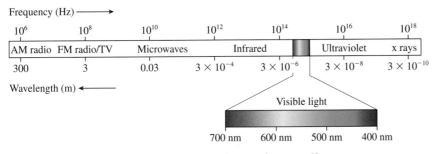

FIGURE 20.25 The electromagnetic spectrum from 10^6 Hz to 10^{18} Hz.

EXAMPLE 20.8 Traveling at the speed of light
A satellite exploring Jupiter transmits data to the earth as a radio wave with a frequency of 200 MHz. What is the wavelength of the electromagnetic wave, and how long does it take the signal to travel 800 million kilometers from Jupiter to the earth?

SOLVE Radio waves are sinusoidal electromagnetic waves that travel with speed c. Thus

$$\lambda = \frac{c}{f} = \frac{3.00 \times 10^8 \text{ m/s}}{2.00 \times 10^8 \text{ Hz}} = 1.5 \text{ m}$$

The time needed to travel 800×10^6 km $= 8.0 \times 10^{11}$ m is

$$\Delta t = \frac{\Delta x}{c} = \frac{8.0 \times 10^{11} \text{ m}}{3.00 \times 10^8 \text{ m/s}} = 2700 \text{ s} = 45 \text{ min}$$

The Index of Refraction

Light waves travel with speed c in a vacuum, but they slow down as they pass through transparent materials such as water or glass or even, to a very slight extent, air. The slowdown is a consequence of interactions between the electromagnetic field of the wave and the electrons in the material. The speed of light in a material is characterized by the material's **index of refraction** n, defined as

$$n = \frac{\text{speed of light in a vacuum}}{\text{speed of light in the material}} = \frac{c}{v} \qquad (20.29)$$

where v is the speed of light in the material. The index of refraction of a material is always greater than 1 because $v < c$. A vacuum has $n = 1$ exactly. Table 20.2 shows the index of refraction for several materials. You can see that liquids and solids have larger indices of refraction than gases.

TABLE 20.2 Typical indices of refraction

Material	Index of refraction
Vacuum	1 exactly
Air	1.0003
Water	1.33
Glass	1.50
Diamond	2.42

NOTE ▶ An accurate value for the index of refraction of air is relevant only in very precise measurements. We will assume $n_{air} = 1.00$ in this text. ◀

If the speed of a light wave changes as it enters into a transparent material, such as glass, what happens to the light's frequency and wavelength? Because $v = \lambda f$, either λ or f or both have to change when v changes.

As an analogy, think of a sound wave in the air as it impinges on the surface of a pool of water. As the air oscillates back and forth, it periodically pushes on the surface of the water. These pushes generate the compressions of the sound wave that continues on into the water. Because each push of the air causes one compression of the water, the frequency of the sound wave in the water must be *exactly the same* as the frequency of the sound wave in the air. In other words, **the frequency of a wave does not change as the wave moves from one medium to another.**

The same is true for electromagnetic waves, although the pushes are a bit more complex as the electric and magnetic fields of the wave interact with the atoms at the surface of the material. Nonetheless, the frequency does not change as the wave moves from one material to another.

Figure 20.26 shows a light wave passing through a transparent material with index of refraction n. As the wave travels through vacuum it has wavelength λ_{vac} and frequency f_{vac} such that $\lambda_{vac} f_{vac} = c$. In the material, $\lambda_{mat} f_{mat} = v = c/n$. The frequency does not change as the wave enters ($f_{mat} = f_{vac}$), so the wavelength must. The wavelength in the material is

$$\lambda_{mat} = \frac{v}{f_{mat}} = \frac{c}{n f_{mat}} = \frac{c}{n f_{vac}} = \frac{\lambda_{vac}}{n} \qquad (20.30)$$

A transparent material in which light travels slower, at speed $v = c/n$

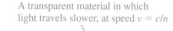

Vacuum $n = 1$ Index n $n = 1$

λ_{vac} $\lambda = \lambda_{vac}/n$

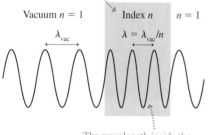

The wavelength inside the material decreases, but the frequency doesn't change.

FIGURE 20.26 Light passing through a transparent material with index of refraction n.

The wavelength in the transparent material is less than the wavelength in vacuum. This makes sense. Suppose a marching band is marching at one step per second at a speed of 1 m/s. Suddenly they slow their speed to $\frac{1}{2}$ m/s but maintain their march at one step per second. The only way to go slower while marching at the same pace is to take *smaller steps*. When a light wave enters a material, the only way it can go slower while oscillating at the same frequency is to have a *smaller wavelength*.

EXAMPLE 20.9 Light traveling through glass
Orange light with a wavelength of 600 nm is incident upon a 1.00-mm-thick glass microscope slide.

a. What is the light speed in the glass?
b. How many wavelengths of the light are inside the slide?

SOLVE
a. From Table 20.2 we see that the index of refraction of glass is $n_{glass} = 1.50$. Thus the speed of light in glass is

$$v_{glass} = \frac{c}{n_{glass}} = \frac{3.00 \times 10^8 \text{ m/s}}{1.50} = 2.00 \times 10^8 \text{ m/s}$$

b. Because $n_{air} = 1.00$, the wavelength of the light is the same in air and vacuum: $\lambda_{vac} = \lambda_{air} = 600$ nm. Thus the wavelength inside the glass is

$$\lambda_{glass} = \frac{\lambda_{vac}}{n_{glass}} = \frac{600 \text{ nm}}{1.50} = 400 \text{ nm} = 4.00 \times 10^{-7} \text{ m}$$

N wavelengths span a distance $d = N\lambda$, so the number of wavelengths in $d = 1.00$ mm is

$$N = \frac{d}{\lambda} = \frac{1.00 \times 10^{-3} \text{ m}}{4.00 \times 10^{-7} \text{ m}} = 2500$$

ASSESS The fact that 2500 wavelengths fit within 1 mm shows how small the wavelengths of light are.

STOP TO THINK 20.5   A light wave travels through three transparent materials of equal thickness. Rank in order, from the largest to smallest, the indices of refraction n_1, n_2, and n_3.

20.6 Power and Intensity

A traveling wave transfers energy from one point to another. The sound wave from a loudspeaker sets your eardrum into motion and, if at the proper frequency, can shatter a glass. Light waves from the sun warm the earth and, if focused with a lens, can start a fire. The *power* of a wave is the rate, in joules per second, at which the wave transfers energy. As you learned in Chapter 11, power is measured in watts. A loudspeaker might emit 2 W of power, meaning that energy in the form of sound waves is radiated at the rate of 2 joules per second. A light bulb might emit 2 W, or 2 J/s, of visible light. (In fact, this is about right for a so-called 100 watt bulb, with the other 98 W of power being emitted as heat, or infrared radiation, rather than as visible light.)

Imagine doing two experiments with a light bulb that emits 2 W of visible light. In the first, you hang the bulb in the center of a room and allow the light to illuminate the walls. In the second experiment, you use mirrors and lenses to "capture" the bulb's light and focus it onto a small spot on one wall. This is what a slide projector does. The energy emitted by the bulb is the same in both cases, but, as you know, the light is much brighter when focused onto a small area. We would say that the focused light is more *intense* than the diffuse light that goes in all directions. Similarly, a loudspeaker that beams its sound forward into a small area produces a louder sound in that area than a speaker of equal power that radiates the sound in all directions. Quantities such as brightness and loudness depend not only on the rate of energy transfer, or power, but also on the *area* that receives that power.

Figure 20.27 shows a wave impinging on a surface of area a. The surface is perpendicular to the direction in which the wave is traveling. This might be a real, physical surface, such as your eardrum or a photovoltaic cell, but it could equally well be a mathematical surface in space that the wave passes right through. If the wave has power P, we define the **intensity** I of the wave to be

$$I = \frac{P}{a} = \text{power-to-area ratio} \tag{20.31}$$

The SI units of intensity are W/m^2. Because intensity is a power-to-area ratio, a wave focused into a small area will have a larger intensity than a wave of equal power that is spread out over a large area.

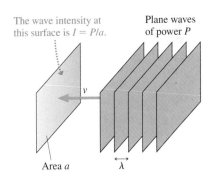

Energy from the sun is a practical and efficient way to heat water, as these solar panels are doing.

The wave intensity at this surface is $I = P/a$.

Plane waves of power P

Area a

FIGURE 20.27 Plane waves of power P impinge on area a with intensity $I = P/a$.

EXAMPLE 20.10   The intensity of a laser beam
A helium-neon laser, the kind that provides the familiar red light of classroom demonstrations and supermarket checkout scanners, emits 1.0 mW of light power into a laser beam that is 1.0 mm in diameter. What is the intensity of the laser beam?

MODEL   The laser beam is a light wave.

SOLVE   The light waves of the laser beam pass through a mathematical surface that is a circle of diameter 1 mm. The intensity of the laser beam is

$$I = \frac{P}{a} = \frac{P}{\pi r^2} = \frac{0.0010 \text{ W}}{\pi (0.00050 \text{ m})^2} = 1270 \text{ W/m}^2$$

ASSESS This is roughly the intensity of sunlight at noon on a summer day. The difference between the sun and a small laser is not their intensities, which are about the same, but their powers. The laser has a small power of 1 mW. It can produce a very intense wave only because the area through which the wave passes is very small. The sun, by contrast, radiates a total power $P_{sun} \approx 4 \times 10^{26}$ W. This immense power is spread through *all* of space, producing an intensity of 1400 W/m² at a distance of 1.5×10^{11} m, the radius of the earth's orbit.

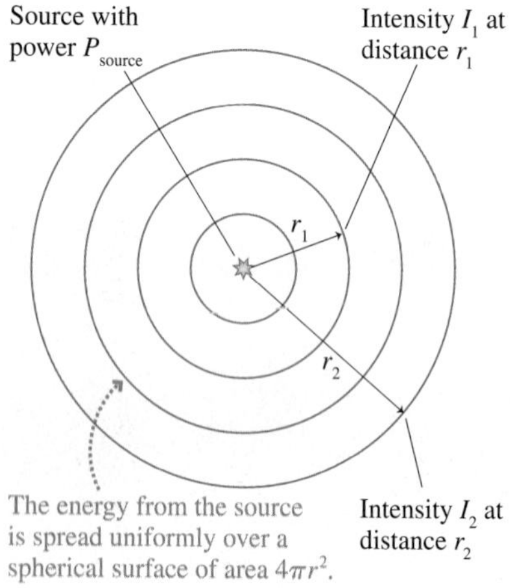

Source with power P_{source} — Intensity I_1 at distance r_1

r_1

r_2

The energy from the source is spread uniformly over a spherical surface of area $4\pi r^2$. — Intensity I_2 at distance r_2

FIGURE 20.28 A source emitting uniform spherical waves.

If a source of spherical waves radiates uniformly in all directions, then, as Figure 20.28 shows, the power at distance r is spread uniformly over the surface of a sphere of radius r. The surface area of a sphere is $a = 4\pi r^2$, so the intensity of a uniform spherical wave is

$$I = \frac{P_{source}}{4\pi r^2} \qquad \text{(intensity of a uniform spherical source)} \qquad (20.32)$$

The inverse-square dependence of r is really just a statement of energy conservation. The source emits energy at the rate P joules per second. The energy is spread over a larger and larger area as the wave moves outward. Consequently, the energy *per unit area* must decrease in proportion to the surface area of a sphere.

If the intensity at distance r_1 is $I_1 = P_{source}/4\pi r_1^2$ and the intensity at r_2 is $I_2 = P_{source}/4\pi r_2^2$, then you can see that the intensity *ratio* is

$$\frac{I_1}{I_2} = \frac{r_2^2}{r_1^2} \qquad (20.33)$$

You can use Equation 20.33 to compare the intensities at two distances from a source without needing to know the power of the source.

NOTE ▶ Wave intensities are strongly affected by reflections and absorption. Equations 20.32 and 20.33 apply to situations such as the light from a star or the sound from a firework exploding high in the air. Indoor sound does *not* obey a simple inverse-square law because of the many reflecting surfaces. ◀

For a sinusoidal wave, each particle in the medium oscillates back and forth in simple harmonic motion. You learned in Chapter 14 that a particle in SHM with amplitude A has energy $E = \frac{1}{2}kA^2$, where k is the spring constant of the medium, *not* the wave number. It is this oscillatory energy of the medium that is transferred, particle to particle, as the wave moves through the medium.

Because a wave's intensity is proportional to the rate at which energy is transferred through the medium, and because the oscillatory energy in the medium is proportional to the *square* of the amplitude, we can infer that for *any* wave

$$I = CA^2 \qquad (20.34)$$

where C is a proportionality constant that depends on the type of wave. That is, **the intensity of a wave is proportional to the square of its amplitude.** If you double the amplitude of a wave, you increase its intensity by a factor of four. This relationship between intensity and amplitude will be important when we study some of the properties of light waves.

20.7 The Doppler Effect

10.8, 10.9

Our final topic for this chapter is an interesting effect that occurs when you are in motion relative to a wave source. It is called the *Doppler effect*. You've likely noticed that the pitch of an ambulance's siren drops as it goes past you. A higher pitch suddenly becomes a lower pitch. Why?

Figure 20.29 shows a source of sound waves moving away from Pablo and toward Nancy at a steady speed v_s. The subscript s indicates that this is the speed of the source, not the speed of the waves. The source is emitting sound waves of frequency f_0 as it travels. Part a of the figure is a motion diagram showing the position of the source times $t = 0$, T, $2T$, and $3T$, where $T = 1/f_0$ is the period of the waves.

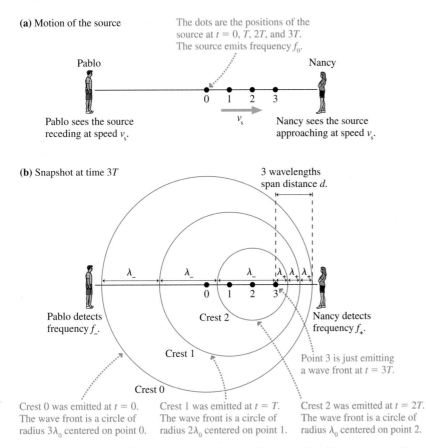

(a) Motion of the source

The dots are the positions of the source at $t = 0$, T, $2T$, and $3T$. The source emits frequency f_0.

Pablo

Nancy

0 1 2 3

v_s

Pablo sees the source receding at speed v_s.

Nancy sees the source approaching at speed v_s.

(b) Snapshot at time $3T$

3 wavelengths span distance d.

λ_- λ_- λ_- λ_+ λ_+ λ_+

0 1 2 3

Pablo detects frequency f_-.

Crest 2

Nancy detects frequency f_+.

Crest 1

Point 3 is just emitting a wave front at $t = 3T$.

Crest 0

Crest 0 was emitted at $t = 0$. The wave front is a circle of radius $3\lambda_0$ centered on point 0.

Crest 1 was emitted at $t = T$. The wave front is a circle of radius $2\lambda_0$ centered on point 1.

Crest 2 was emitted at $t = 2T$. The wave front is a circle of radius λ_0 centered on point 2.

FIGURE 20.29 A motion diagram showing the wave fronts emitted by a source as it moves to the right at speed v_s.

Nancy measures the frequency of the wave emitted by the *approaching source* to be f_+. At the same time, Pablo measures the frequency of the wave emitted by the *receding source* to be f_-. Our task is to relate f_+ and f_- to the source frequency f_0 and speed v_s.

After a wave crest leaves the source, its motion is governed by the properties of the medium. That is, the motion of the source cannot affect a wave that has already been emitted. Thus each circular wave front in Figure 20.29b is centered on the point from which it was emitted. The wave crest from point 3 was emitted just as this figure was made, but it hasn't yet had time to travel any distance.

You can see that the wave crests are bunched up in the direction the source is moving, stretched out behind it. The distance between one crest and the next is one wavelength, so the wavelength λ_+ that Nancy measures is *less* than the wavelength $\lambda_0 = v/f_0$ that would be emitted if the source were at rest. Similarly, λ_- behind the source is larger than λ_0.

These crests move through the medium at the wave speed v. Consequently, the frequency $f_+ = v/\lambda_+$ detected by the observer whom the source is approaching is *higher* than the frequency f_0 emitted by the source. Similarly, $f_- = v/\lambda_-$

detected behind the source is *lower* than frequency f_0. This change of frequency when a source moves relative to an observer is called the **Doppler effect.**

The wavelength detected by Nancy is $\lambda_+ = d/3$, where d is the difference between how far the wave has moved and how far the source has moved at time $t = 3T$. These distances are

$$\Delta x_{\text{wave}} = vt = 3vT$$
$$\Delta x_{\text{source}} = v_s t = 3v_s T \tag{20.35}$$

Thus the wavelength of the wave emitted by an approaching source is

$$\lambda_+ = \frac{d}{3} = \frac{\Delta x_{\text{wave}} - \Delta x_{\text{source}}}{3} = \frac{3vT - 3v_s T}{3} = (v - v_s)T \tag{20.36}$$

You can see that our arbitrary choice of three periods was not relevant because the 3 cancels. Thus the frequency detected in Nancy's direction is

$$f_+ = \frac{v}{\lambda_+} = \frac{v}{(v - v_s)T} = \frac{v}{(v - v_s)} f_0 \tag{20.37}$$

where $f_0 = 1/T$ is the frequency of the source and is the frequency you would detect if the source were at rest. We'll find it convenient to write the detected frequency as

$$f_+ = \frac{f_0}{1 - v_s/v} \qquad \text{(Doppler effect for an approaching source)}$$
$$f_- = \frac{f_0}{1 + v_s/v} \qquad \text{(Doppler effect for a receding source)} \tag{20.38}$$

Proof of the second version, for the frequency f_- of a receding source, will be left for a homework problem. You can see that $f_+ > f_0$ in front of the source, because the denominator is less than 1, and $f_- < f_0$ behind the source.

EXAMPLE 20.11 **How fast are the police traveling?**

A police siren has a frequency of 550 Hz as the police car approaches you, 450 Hz after it has passed you and is receding. How fast are the police traveling? The temperature is 20°C.

MODEL The siren's frequency is altered by the Doppler effect. The frequency is f_+ as the car approaches and f_- as it moves away.

SOLVE To find v_s, rewrite Equations 20.38 as

$$f_0 = (1 + v_s/v)f_-$$
$$f_0 = (1 - v_s/v)f_+$$

Subtract the second equation from the first, giving

$$0 = f_- - f_+ + \frac{v_s}{v}(f_- + f_+)$$

This is easily solved to give

$$v_s = \frac{f_+ - f_-}{f_+ + f_-} v = \frac{100 \text{ Hz}}{1000 \text{ Hz}} 343 \text{ m/s} = 34.3 \text{ m/s}$$

ASSESS If you now solve for the siren frequency when at rest, you will find $f_0 = 495$ Hz. Surprisingly, the at-rest frequency is *not* halfway between f_- and f_+.

NOTE ▶ The frequency of an approaching source is shifted upward, from f_0 to f_+, but the frequency *does not change* as the source gets closer. It's often said that the frequency *rises* as a source approaches, but you can see that is not the case. What does rise is the intensity, or loudness, of the sound. Interestingly, a sound of constant frequency but increasing loudness is often *perceived* to be increasing in pitch. You might perceive that the pitch of an approaching ambulance is rising, but measurements would show that the frequency remains constant as the intensity increases. ◀

A Stationary Source and a Moving Observer

Suppose the police car in Example 20.11 is at rest while you drive toward it at 34.3 m/s. You might think that this is equivalent to having the police car move toward you at 34.3 m/s, but it isn't. Mechanical waves move through a medium, and the Doppler effect depends not just on how the source and the observer move with respect to each other but also how they move with respect to the medium. We'll omit the proof, but it's not hard to show that the frequencies heard by an observer moving at speed v_o relative to a stationary source emitting frequency f_0 are

$$f_+ = (1 + v_o/v)f_0 \quad \text{(observer approaching a source)}$$

$$f_- = (1 - v_o/v)f_0 \quad \text{(observer receding from a source)}$$

(20.39)

A quick calculation shows that the frequency of the police siren as you approach it at 34.3 m/s is 545 Hz, not the 550 Hz you heard as it approached you at 34.3 m/s.

The Doppler Effect for Light Waves

The Doppler effect is observed for all types of waves, not just sound waves. If a source of light waves is receding from you, the wavelength λ_- that you detect is longer than the wavelength λ_0 emitted by the source. Because the wavelength is shifted toward the red end of the visible spectrum, the longer wavelengths of light, this effect is called the **red shift.** Similarly, the light you detect from a source moving toward you is **blue shifted** to shorter wavelengths.

Although the underlying reason for the Doppler shift for light is the same as for sound waves, there is one fundamental difference. We derived Equation 20.38 for the Doppler-shifted frequencies by measuring the wave speed v relative to the medium. For electromagnetic waves in empty space, there is no medium. Consequently, we would need to turn to Einstein's theory of relativity to determine the frequency of light waves from a moving source. The result, which we will state without proof, is

$$\lambda_{\text{red}} = \sqrt{\frac{1 + v_s/c}{1 - v_s/c}}\, \lambda_0$$

(Doppler effect for the light of a receding source)

(20.40)

$$\lambda_{\text{blue}} = \sqrt{\frac{1 - v_s/c}{1 + v_s/c}}\, \lambda_0$$

(Doppler effect for the light of an approaching source)

Here v_s is the speed of the source *relative to* the observer.

EXAMPLE 20.12 **Measuring the velocity of a galaxy**

Hydrogen atoms in the laboratory emit red light with wavelength 656 nm. In the light from a distant galaxy, this "spectral line" is observed at 691 nm. What is the speed of this galaxy relative to the earth?

MODEL The observed wavelength is longer than the wavelength emitted by atoms at rest with respect to the observer (i.e., red shifted), so we are looking at light emitted from a galaxy that is receding from us.

SOLVE Squaring the expression for λ_{red} in Equation 20.40 and solving for v_s gives

$$v_s = \frac{(\lambda_{\text{red}}/\lambda_0)^2 - 1}{(\lambda_{\text{red}}/\lambda_0)^2 + 1}\, c$$

$$= \frac{(691 \text{ nm}/656 \text{ nm})^2 - 1}{(691 \text{ nm}/656 \text{ nm})^2 + 1}\, c$$

$$= 0.052c = 1.56 \times 10^7 \text{ m/s}$$

ASSESS The galaxy is moving away from the earth at about 5% of the speed of light!

FIGURE 20.30 A Hubble Space Telescope picture of a quasar.

In the 1920s, an analysis of the red shifts of many galaxies led the astronomer Edwin Hubble to the conclusion that the galaxies of the universe are *all* moving apart from each other. Extrapolating backward in time must bring us to a point when all the matter of the universe—and even space itself, according to the theory of relativity—began rushing out of a primordial fireball. Many observations and measurements since have given support to the idea that the universe began in a *Big Bang* about 15 billion years ago.

As an example, Figure 20.30 is a Hubble Space Telescope picture of a *quasar,* short for *quasistellar object.* Quasars are extraordinarily powerful sources of light and radio waves. The light reaching us from quasars is highly red shifted, corresponding in some cases to objects that are moving away from us at greater than 90% of the speed of light. Astronomers have determined that some quasars are 10 to 12 *billion* light years away from the earth, hence the light we see was emitted when the universe was only about 25% of its present age. Today, the red shifts of distant quasars and supernovae (exploding stars) are being used to refine our understanding of the structure and evolution of the universe.

STOP TO THINK 20.6 Amy and Zack are both listening to the source of sound waves that is moving to the right. Compare the frequencies each hears.

a. $f_{Amy} > f_{Zack}$

b. $f_{Amy} = f_{Zack}$

c. $f_{Amy} < f_{Zack}$

SUMMARY

The goal of Chapter 20 has been to learn the basic properties of traveling waves.

GENERAL PRINCIPLES

The Wave Model

This model is based on the idea of a **traveling wave**, which is an organized disturbance traveling at a well-defined **wave speed** v.

- In transverse waves the particles of the medium move perpendicular to the direction in which the wave travels.

- In longitudinal waves the particles of the medium move parallel to the direction in which the wave travels.

A wave transfers **energy,** but no material or substance is transferred outward from the source.

Three basic types of waves:

- **Mechanical waves** travel through a material medium such as water or air.

- **Electromagnetic waves** require no material medium and can travel through a vacuum.

- **Matter waves** describe the wavelike characteristics of atomic-level particles.

For mechanical waves, the speed of the wave is a property of the medium. Speed does not depend on the size or shape of the wave.

IMPORTANT CONCEPTS

The **displacement** D of a wave is a function of both position (where) and time (when).

- A **snapshot graph** shows the wave's displacement as a function of position at a single instant of time.

- A **history graph** shows the wave's displacement as a function of time at a single point in space.

A wave traveling in the positive x-direction with speed v must be a function of the form $D(x - vt)$.

A wave traveling in the negative x-direction with speed v must be a function of the form $D(x + vt)$.

Sinusoidal waves are periodic in both time (period T) and space (wavelength λ).

$$D(x, t) = A \sin\left[2\pi(x/\lambda - t/T) + \phi_0\right]$$
$$= A \sin(kx - \omega t + \phi_0)$$

where A is the **amplitude,** $k = 2\pi/\lambda$ is the **wave number,** $\omega = 2\pi f = 2\pi/T$ is the **angular frequency,** and ϕ_0 is the **phase constant** that describes initial conditions.

One-dimensional waves   Two- and three-dimensional waves

The fundamental relationship for any sinusoidal wave is $v = \lambda f$.

APPLICATIONS

Wave speeds for some specific waves:

- **String** (transverse): $v = \sqrt{T_s/\mu}$

- **Sound** (longitudinal): $v = 343$ m/s in 20°C air

- **Light** (transverse): $v = c/n$, where $c = 3.00 \times 10^8$ m/s is the speed of light in a vacuum and n is the material's **index of refraction.**

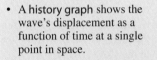

The wave **intensity** is the power-to-area ratio

$$I = P/A$$

For a circular or spherical wave

$$I = P_{\text{source}}/4\pi r^2$$

The **Doppler effect** occurs when a wave source and detector are moving with respect to each other: the frequency detected differs from the frequency f_0 emitted.

Approaching source

$$f_+ = \frac{f_0}{1 - v_s/v}$$

Receding source

$$f_- = \frac{f_0}{1 + v_s/v}$$

Observer approaching a source

$$f_+ = (1 + v_o/v)f_0$$

Observer receding from a source

$$f_- = (1 - v_o/v)f_0$$

The Doppler effect for light uses a result derived from the theory of relativity.

TERMS AND NOTATION

wave model	trailing edge	plane wave
traveling wave	linear density, μ	phase, ϕ
mechanical waves	snapshot graph	compression
electromagnetic waves	history graph	rarefaction
matter waves	sinusoidal wave	electromagnetic spectrum
medium	amplitude, A	index of refraction, n
disturbance	wavelength, λ	intensity, I
wave speed, v	wave number, k	Doppler effect
transverse wave	wave front	red shift
longitudinal wave	circular wave	blue shift
leading edge	spherical wave	

EXERCISES AND PROBLEMS

Exercises

Section 20.2 One-Dimensional Waves

1. Draw the history graph $D(x = 6 \text{ m}, t)$ of this wave at $x = 6$ m.

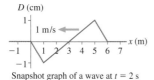

FIGURE EX20.1 Snapshot graph of a wave at $t = 0$ s

2. Draw the history graph $D(x = 0 \text{ m}, t)$ of this wave at $x = 0$ m.

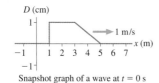

FIGURE EX20.2 Snapshot graph of a wave at $t = 2$ s

3. Draw the snapshot graph $D(x, t = 1 \text{ s})$ of this wave at $t = 1$ s.

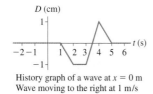

History graph of a wave at $x = 0$ m
FIGURE EX20.3 Wave moving to the right at 1 m/s

4. Draw the snapshot graph $D(x, t = 0 \text{ s})$ of this wave at $t = 0$ s.

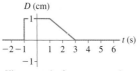

History graph of a wave at $x = 2$ m
FIGURE EX20.4 Wave moving to the left at 1 m/s

5. Figure Ex20.5 is the snapshot graph at $t = 0$ s of a *longitudinal* wave. Draw the corresponding picture of the particle positions, as was done in Figure 20.10. Let the equilibrium spacing between the particles be 1.0 cm.

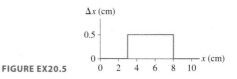

FIGURE EX20.5

6. Figure Ex20.6 is a picture at $t = 0$ s of the particles in a medium as a longitudinal wave is passing through. The equilibrium spacing between the particles is 1.0 cm. Draw the snapshot graph $D(x, t = 0 \text{ s})$ of this wave at $t = 0$ s.

FIGURE EX20.6

7. The wave speed on a string under tension is 200 m/s. What is the speed if the tension is doubled?
8. The wave speed on a string is 150 m/s when the tension is 75 N. What tension will give a speed of 180 m/s?
9. A wave travels along a string at a speed of 280 m/s. What will be the speed if the string is replaced by one made of the same material and under the same tension but having twice the radius?

Section 20.3 Sinusoidal Waves

10. A wave has angular frequency 30 rad/s and wavelength 2.0 m. What are its (a) wave number and (b) wave speed?
11. A wave travels with speed 200 m/s. Its wave number is 1.5 rad/m. What are its (a) wavelength and (b) frequency?
12. The displacement of a wave traveling in the positive x-direction is $D(x, t) = (3.5 \text{ cm}) \sin(2.7x - 124t)$, where x is in m and t is in s. What are the (a) frequency, (b) wavelength, and (c) speed of this wave?

13. The displacement of a wave traveling in the negative y-direction is $D(y, t) = (5.2 \text{ cm}) \sin(5.5y + 72t)$, where y is in m and t is in s. What are the (a) frequency, (b) wavelength, and (c) speed of this wave?

14. Figure Ex20.14 is a snapshot graph at $t = \frac{1}{4}$ s of a traveling wave with displacement $D(x, t) = (2.0 \text{ cm}) \sin(2\pi x - 4\pi t)$, where x is in m and t is in s.

    a. Reproduce this graph on your page, then use a *dotted* line to show the snapshot graph at $t = 0$ s and a *dashed* line to show the snapshot graph at $t = \frac{1}{8}$ s.

    b. What is the speed of this wave?

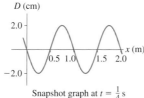

FIGURE EX20.14   Snapshot graph at $t = \frac{1}{4}$ s

15. What are the amplitude, wavelength, and frequency of this wave?

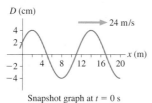

FIGURE EX20.15   Snapshot graph at $t = 0$ s

16. What are the amplitude, frequency, and wavelength of this wave?

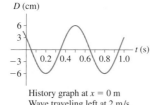

FIGURE EX20.16   History graph at $x = 0$ m
Wave traveling left at 2 m/s

Section 20.4 Waves in Two and Three Dimensions

17. A circular wave travels outward from the origin. At one instant of time, the phase at $r_1 = 20$ cm is 0 rad and the phase at $r_2 = 80$ cm is 3π rad. What is the wavelength of the wave?

18. A spherical wave with a wavelength of 2.0 m is emitted from the origin. At one instant of time, the phase at $r = 4.0$ m is π rad. At that instant, what is the phase at $r = 3.5$ m and at $r = 4.5$ m?

19. A loudspeaker at the origin emits sound waves on a day when the speed of sound is 340 m/s. A crest of the wave simultaneously passes listeners at the (x, y) coordinates (40 m, 0 m) and (0 m, 30 m). What are the lowest two possible frequencies of the sound?

20. A sound source is located somewhere along the x-axis. Experiments show that the same wave front simultaneously reaches listeners at $x = -7.0$ m and $x = +3.0$ m.

    a. What is the x-coordinate of the source?

    b. A third listener is positioned along the positive y-axis. What is her y-coordinate if the same wave front reaches her at the same instant it does the first two listeners?

Section 20.5 Sound and Light

21. The back wall of an auditorium is 26 m from the stage. If you are seated in the middle row, how much time elapses between a sound from the stage reaching your ear directly and the same sound reaching your ear after reflecting from the back wall?

22. A hammer taps on the end of a 4.0-m-long metal bar at room temperature. A microphone at the other end of the bar picks up two pulses of sound, one that travels through the metal and one that travels through the air. The pulses are separated in time by 11.0 ms. What is the speed of sound in this metal?

23. Oil explorers set off explosives to make loud sounds, then listen for the echoes from underground oil deposits. Geologists suspect that there is oil under 500-m-deep Lake Physics. It's known that Lake Physics is carved out of a granite basin. Explorers detect a weak echo 0.94 s after exploding dynamite at the lake surface. If it's really oil, how deep will they have to drill into the granite to reach it?

24. a. What is the wavelength of a 2.0 MHz ultrasound wave traveling through aluminum?

    b. What frequency of electromagnetic wave would have the same wavelength as the ultrasound wave of part a?

25. a. At 20°C, what is the frequency of a sound wave in air with a wavelength of 20 cm?

    b. What is the frequency of an electromagnetic wave with a wavelength of 20 cm?

    c. What would be the wavelength of a sound wave in water that has the same frequency as the electromagnetic wave of part b?

26. a. What is the frequency of blue light that has a wavelength of 450 nm?

    b. What is the frequency of red light that has a wavelength of 650 nm?

    c. What is the index of refraction of a material in which the red-light wavelength is 450 nm?

27. a. Telephone signals are often transmitted over long distances by microwaves. What is the frequency of microwave radiation with a wavelength of 3.0 cm?

    b. Microwave signals are beamed between two mountaintops 50 km apart. How long does it take a signal to travel from one mountaintop to the other?

28. a. An FM radio station broadcasts at a frequency of 101.3 MHz. What is the wavelength?

    b. What is the frequency of a sound source that produces the same wavelength in 20°C air?

29. a. How long does it take light to travel through a 3.0-mm-thick piece of window glass?

    b. Through what thickness of water could light travel in the same amount of time?

30. A light wave has a 670 nm wavelength in air. Its wavelength in a transparent solid is 420 nm.

    a. What is the speed of light in this solid?

    b. What is the light's frequency in the solid?

31. A helium-neon laser beam has a wavelength in air of 633 nm. It takes 1.38 ns for the light to travel through 30 cm of an unknown liquid. What is the wavelength of the laser beam in the liquid?

Section 20.6 Power and Intensity

32. A sound wave with intensity 2.0×10^{-3} W/m² is perceived to be modestly loud. Your eardrum is 6.0 mm in diameter. How much energy will be transferred to your eardrum while listening to this sound for 1.0 min?

33. The intensity of electromagnetic waves from the sun is 1.4 kW/m² just above the earth's atmosphere. Eighty percent of this reaches the surface at noon on a clear summer day. Suppose you think of your back as a 30 cm × 50 cm rectangle. How many joules of solar energy fall on your back as you work on your tan for 1.0 hr?

34. The sound intensity from a jack hammer breaking concrete is 2.0 W/m² at a distance of 2.0 m from the point of impact. This is sufficiently loud to cause permanent hearing damage if the operator doesn't wear ear protection. What is the sound intensity for a person watching from 50 m away?

35. A concert loudspeaker suspended high off the ground emits 35 W of sound power. A small microphone with a 1.0 cm² area is 50 m from the speaker.
    a. What is the sound intensity at the position of the microphone?
    b. How much sound energy impinges on the microphone each second?

Section 20.7 The Doppler Effect

36. An opera singer in a convertible sings a note at 600 Hz while cruising down the highway at 90 km/hr. What is the frequency heard by
    a. A person standing beside the road in front of the car?
    b. A person on the ground behind the car?

37. A mother hawk screeches as she dives at you. You recall from biology that female hawks screech at 800 Hz, but you hear the screech at 900 Hz. How fast is the hawk approaching?

38. A whistle you use to call your hunting dog has a frequency of 21 kHz, but your dog is ignoring it. You suspect the whistle may not be working, but you can't hear sounds above 20 kHz. To test it, you ask a friend to blow the whistle, then you hop on your bicycle. In which direction should you ride (toward or away from your friend) and at what minimum speed to know if the whistle is working?

39. A friend of yours is loudly singing a single note at 400 Hz while racing toward you at 25.0 m/s on a day when the speed of sound is 340 m/s.
    a. What frequency do you hear?
    b. What frequency does your friend hear if you suddenly start singing at 400 Hz?

Problems

40. The displacement of a traveling wave is
$$D(x, t) = \begin{cases} 1 \text{ cm} & \text{if } |x - 3t| \leq 1 \\ 0 \text{ cm} & \text{if } |x - 3t| > 1 \end{cases}$$
    where x is in m and t in s.
    a. Draw displacement-versus-position graphs at 1 s intervals from $t = 0$ s to $t = 3$ s. Use an x-axis that goes from −2 to 12 m. Stack the four graphs vertically, similar to Figure 20.12.

    b. Determine the wave speed from the graphs. Explain how you did so.
    c. Determine the wave speed from the equation for $D(x, t)$. Does it agree with your answer to part b?

41. The displacement of a traveling wave is
$$D(x, t) = \begin{cases} 1 \text{ cm} & \text{if } |x + 2t| \leq \frac{1}{2} \\ 0 \text{ cm} & \text{if } |x + 2t| > \frac{1}{2} \end{cases}$$
    where x is in m and t in s.
    a. Draw displacement-versus-position graphs at 1 s intervals from $t = -2$ s to $t = +2$ s. Use an x-axis that goes from −6 to 6 m. Stack the five graphs vertically, similar to Figure 20.12.
    b. Determine the wave speed from the graphs. Explain how you did so.
    c. Determine the wave speed from the equation for $D(x, t)$. Does it agree with your answer to part b?

42. This is a snapshot graph at $t = 0$ s of a 5.0 Hz wave traveling to the left.
    a. What is the wave speed?
    b. What is the phase constant of the wave?
    c. Write the displacement equation for this wave.

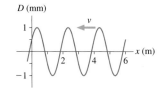

FIGURE P20.42   Snapshot graph at $t = 0$ s

43. This is a history graph at $x = 0$ m of a wave traveling in the positive x-direction at 4.0 m/s.
    a. What is the wavelength?
    b. What is the phase constant of the wave?
    c. Write the displacement equation for this wave.

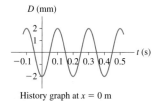

FIGURE P20.43   History graph at $x = 0$ m
Wave traveling right at 4.0 m/s

44. An ultrasound unit sends a 2.4 MHz sound wave into a 25-cm-long tube filled with an unknown liquid. A small microphone right next to the ultrasonic generator detects both the transmitted wave and the sound wave that has reflected off the far end of the tube. The two sound pulses are 4.4 divisions apart on an oscilloscope for which the horizontal time sweep is set to 100 μs/division. What is the speed of sound in the liquid?

45. A 2.0-m-long string is under 20 N of tension. A pulse travels the length of the string in 50 ms. What is the mass of the string?

46. Andy (mass 80 kg) uses a 3.0-m-long rope to pull Bob (mass 60 kg) across the floor ($\mu_k = 0.20$) at a constant speed of 1.0 m/s. Bob signals to Andy to stop by "plucking" the rope, sending a wave pulse forward along the rope. The pulse reaches Andy 150 ms later. What is the mass of the rope?

47. String 1 in Figure P20.47 has linear density 2.0 g/m and string 2 has linear density 4.0 g/m. A student sends pulses in both

directions by quickly pulling up on the knot, then releasing it. What should the string lengths L_1 and L_2 be if the pulses are to reach the ends of the strings simultaneously?

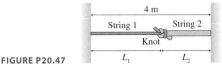

FIGURE P20.47

48. Ships measure the distance to the ocean bottom with *sonar*. A pulse of sound waves is aimed at the ocean bottom, then sensitive microphones listen for the echo. The graph shows the delay time as a function of the ship's position as it crosses 60 km of ocean. Draw a graph of the ocean bottom. Let the ocean surface define $y = 0$ and ocean bottom have negative values of y. This way your graph will be a picture of the ocean bottom. The speed of sound in ocean water varies slightly with temperature, but you can use 1500 m/s as an average value.

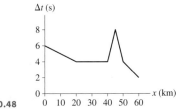

FIGURE P20.48

49. A long string is shaken at its left end at 5.0 Hz, sending a sinusoidal wave down the string at 2.0 m/s. After the wave has traveled 1.0 m, the linear density of the string suddenly decreases by a factor of four. Draw a snapshot graph of the first 3.0 m of the string at an instant when the left end is at its equilibrium position.

50. One cue your hearing system uses to localize a sound (i.e., to tell where a sound is coming from) is the slight difference in the arrival times of the sound at your ears. Your ears are spaced approximately 20 cm apart. Consider a sound source 5.0 m from the center of your head along a line 45° to your right. What is the *difference* in arrival times? Give your answer in microseconds.

Hint: You are looking for the difference between two numbers that are nearly the same. What does this near equality imply about the necessary precision during intermediate stages of the calculation?

51. A 256 Hz sound wave in 20°C air propagates into the water of a swimming pool. What are the water-to-air ratios of the wave's frequency, wave speed, and wave length?

52. Earthquakes are essentially sound waves traveling through the earth. They are called seismic waves. Because the earth is solid, it can support both longitudinal and transverse seismic waves. These travel at different speeds. The speed of longitudinal waves, called P waves, is 8000 m/s. Transverse waves, called S waves, travel at a slower 4500 m/s. A seismograph records the two waves from a distant earthquake. If the S wave arrives 2.0 min after the P wave, how far away was the earthquake? You can assume that the waves travel in straight lines, although actual seismic waves follow more complex routes.

53. One way to monitor global warming is to measure the average temperature of the ocean. Researchers are doing this by mea-

suring the time it takes sound pulses to travel underwater over large distances. At a depth of 1000 m, where ocean temperatures hold steady near 4°C, the average sound speed is 1480 m/s. It's known from laboratory measurements that the sound speed increases 4.0 m/s for every 1.0°C increase in temperature. In one experiment, where sounds generated near California are detected in the South Pacific, the sound waves travel 8000 km. If the smallest time change that can be reliably detected is 1.0 s, what is the smallest change in average temperature that can be measured?

54. A sound wave is described by $D(y, t) = (0.020 \text{ mm}) \times \sin[(8.96 \text{ rad/m})y + (3140 \text{ rad/s})t + \pi/4 \text{ rad}]$, where y is in m and t is in s.
   a. In what direction is this wave traveling?
   b. Along which axis is the air oscillating?
   c. What are the wavelength, the wave speed, and the period of oscillation?
   d. Draw a displacement-versus-time graph $D(y = 1 \text{ m}, t)$ at $y = 1 \text{ m}$ from $t = 0$ s to $t = 0.004$ s.

55. A wave on a string is described by $D(x, t) = (3.0 \text{ cm}) \times \sin[2\pi(x/(2.4 \text{ m}) + t/(0.20 \text{ s}) + 1)]$, where x is in m and t is in s.
   a. In what direction is this wave traveling?
   b. What are the wave speed, the frequency, and the wave number?
   c. At $t = 0.50$ s, what is the displacement of the string at $x = 0.20$ m?

56. A wave on a string is described by $D(x, t) = (2.0 \text{ cm}) \times \sin[(12.57 \text{ rad/m})x - (638 \text{ rad/s})t]$, where x is in m and t is in s. The linear density of the string is 5.0 g/m. What are
   a. The string tension?
   b. The maximum displacement of a point on the string?
   c. The maximum speed of a point on the string?

57. Write the displacement equation for a sinusoidal wave that is traveling in the negative y-direction with wavelength 50 cm, speed 4.0 m/s, and amplitude 5.0 cm. Assume $\phi_0 = 0$.

58. Write the displacement equation for a sinusoidal wave that is traveling in the positive x-direction with frequency 200 Hz, speed 400 m/s, amplitude 0.010 mm, and phase constant $\pi/2$.

59. Show that a wave traveling in the *negative* x-direction with wave speed v must be a function of the form $D(x + vt)$.

60. Show that $D(x, t + T) = D(x, t)$ for a sinusoidal traveling wave. This shows that the wave is periodic with period T.

61. A spherical sound source at the origin emits a sound wave with frequency 13,100 Hz and wave speed 346 m/s. What is the phase difference in degrees and in rad between the two points with (x, y, z) coordinates (1.0 cm, 3.0 cm, 2.0 cm) and $(-1.0 \text{ cm}, 1.5 \text{ cm}, 2.5 \text{ cm})$?

62. A string with linear density 2.0 g/m is stretched along the positive x-axis with tension 20 N. One end of the string, at $x = 0$ m, is tied to a hook that oscillates up and down at a frequency of 100 Hz with a maximum displacement of 1.0 mm. At $t = 0$ s, the hook is at its lowest point.
   a. What are the wave speed on the string and the wavelength?
   b. What are the amplitude and phase constant of the wave?
   c. Write the equation for the displacement $D(x, t)$ of the traveling wave.
   d. What is the string's displacement at $x = 0.50$ m and $t = 15$ ms?

63. A sound wave of frequency 686 Hz is traveling in 20°C air as a plane wave in the positive y-direction. At time $t = 0$ s, one crest of the wave is located in the plane $y = 12.5$ cm.
    a. Draw a graph of $D(y, t = 0 \text{ s})$ as a function of y from $y = -150$ cm to $y = +150$ cm.
    b. What is the phase constant for this wave? There are *two* values that satisfy the mathematics, so justify your answer.
    c. Write the wave equation $D(y, t)$ for the wave, leaving the amplitude A unspecified.
    d. Draw an xy-coordinate system in which y ranges from -150 cm to $+150$ cm. Show the wave fronts of the wave at time $t = 0$ s. Number each of the wave fronts $(1, 2, \ldots)$ at the side of the graph.
    e. Repeat part d at time $t = 0.729$ ms. Keep the same number attached to each wave front so that you can see how the wave fronts have moved.
    f. At $t = 0.729$ ms, what is the phase of the wave at $y = +12.5$ cm and at $y = -12.5$ cm?
64. Figure P20.64 shows a snapshot graph of a wave traveling to the right along a string at 45 m/s. At this instant, what is the velocity of points 1, 2, and 3 on the string?

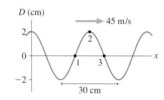

FIGURE P20.64

65. Figure P20.65 is a snapshot graph of the instantaneous *velocity* v_{particle} of the particles on a string. The wave is moving to the left at 50 cm/s. Draw a snapshot graph of the string's displacement at this instant of time.

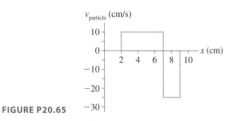

FIGURE P20.65

66. A string that is under 50.0 N of tension has linear density 5.0 g/m. A sinusoidal wave with amplitude 3.0 cm and wavelength 2.0 m travels along the string. What is the maximum velocity of a particle on the string?
67. A sinusoidal wave travels along a stretched string. A particle on the string has a maximum velocity of 2.0 m/s and a maximum acceleration of 200 m/s². What are the frequency and amplitude of the wave?
68. The sun emits electromagnetic waves with a power of 4×10^{26} W. Determine the intensity of electromagnetic waves from the sun just outside the atmospheres of Venus, the earth, and Mars.
69. a. A 100 W lightbulb produces 2.0 W of visible light. (The other 98 W are dissipated as heat and infrared radiation.) What is the light intensity on a wall 2.0 m away from the lightbulb?
    b. A krypton laser produces a cylindrical red laser beam 2.0 mm in diameter with 2.0 W of power. What is the light intensity on a wall 2.0 m away from the laser?

70. An AM radio station broadcasts with a power of 25 kW at a frequency of 920 kHz. Estimate the intensity of the radio wave at a point 10 km from the broadcast antenna.
71. The sound intensity 50 m from a wailing tornado siren is 0.10 W/m².
    a. What is the intensity at 1000 m?
    b. The weakest intensity likely to be heard over background noise is $\approx 1 \ \mu$W/m². Estimate the maximum distance at which the siren can be heard.
72. Lasers can be used to drill or cut material. One such laser generates a series of high-intensity pulses rather than a continuous beam of light. Each pulse contains 500 mJ of energy and lasts 10 ns. The laser fires 10 such pulses per second.
    a. What is the *peak power* of the laser light? The peak power is the power output during one of the 10 ns pulses.
    b. What is the average power output of the laser? The average power is the total energy delivered per second.
    c. A lens focuses the laser beam to a 10-μm-diameter circle on the target. During a pulse, what is the light intensity on the target?
    d. The intensity of sunlight at midday is about 1100 W/m². What is the ratio of the laser intensity on the target to the intensity of the midday sun?
73. A bat locates insects by emitting ultrasonic "chirps" and then listening for echoes from the bugs. Suppose a bat chirp has a frequency of 25 kHz. How fast would the bat have to fly, and in what direction, for you to just barely be able to hear the chirp at 20 kHz?
74. A physics professor demonstrates the Doppler effect by tying a 600 Hz sound generator to a 1.0-m-long rope and whirling it around her head in a horizontal circle at 100 rpm. What are the highest and lowest frequencies heard by a student in the classroom?
75. Show that the Doppler frequency f_- of a receding source is $f_- = f_0/(1 + v_s/v)$.
76. A starship approaches its home planet at a speed of $0.1c$. When it is 54×10^6 km away, it uses its green laser beam $(\lambda = 540$ nm) to signal its approach.
    a. How long does the signal take to travel to the home planet?
    b. At what wavelength is the signal detected on the home planet?
77. You are cruising to Jupiter at the posted speed limit of $0.1c$ when suddenly a daredevil passes you, going in the same direction, at $0.3c$. At what wavelength does your rocket cruiser's light detector "see" his red tail lights? Is this wavelength ultraviolet, visible, or infrared? Use 650 nm for the wavelength of red light.
78. Wavelengths of light from a distant galaxy are found to be 0.5% longer than the corresponding wavelengths measured in a terrestrial laboratory. Is the galaxy approaching or receding from the earth? At what speed?
79. You have just been pulled over for running a red light, and the police officer has informed you that the fine will be $250. In desperation, you suddenly recall an idea that your physics professor recently discussed in class. In your calmest voice, you tell the officer that the laws of physics prevented you from knowing that the light was red. In fact, as you drove toward it, the light was Doppler shifted to where it appeared green to you. "OK," says the officer, "Then I'll ticket you for speeding. The fine is $1 for every 1 km/hr over the posted speed limit of 50 km/hr." How big is your fine? Use 650 nm as the wavelength of red light and 540 nm as the wavelength of green light.

Challenge Problems

80. Figure CP20.80 shows two masses hanging from a steel wire. The mass of the wire is 60.0 g. A wave pulse travels along the wire from point 1 to point 2 in 24.0 ms. What is mass m?

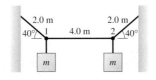

FIGURE CP20.80

81. A wire is made by welding together two metals having different densities. Figure CP20.81 shows a 2.0-m-long section of wire centered on the junction, but the wire extends much farther in both directions. The wire is placed under 2250 N tension, then a 1500 Hz wave with an amplitude of 3.0 mm is sent down the wire. How many wavelengths (complete cycles) of the wave are in this 2.0-m-long section of the wire?

FIGURE CP20.81

82. A rope of mass m and length L hangs from a ceiling.
    a. Show that the wave speed on the rope a distance y above the lower end is $v = \sqrt{gy}$.
    b. Show that the time for a pulse to travel the length of the string is $\Delta t = 2\sqrt{L/g}$.

83. Some modern optical devices are made with glass whose index of refraction changes with distance from the front surface. Figure CP20.83 shows the index of refraction as a function of the distance into a slab of glass of thickness L. The index of refraction increases linearly from n_1 at the front surface to n_2 at the rear surface.
    a. Find an expression for the time light takes to travel through this piece of glass.
    b. Evaluate your expression for a 1.0-cm-thick piece of glass for which $n_1 = 1.50$ and $n_2 = 1.60$.

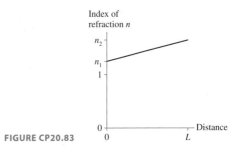

FIGURE CP20.83

STOP TO THINK ANSWERS

Stop to Think 20.1: b. The wave is traveling to the right at 2 m/s, so each point on the wave passed $x = 0$ m, the point of interest, 2 s before reaching $x = 4$ m. The graph has the same shape, but everything happens 2 s earlier.

Stop to Think 20.2: d and e. The wave speed depends on properties of the medium, not on how you generate the wave. For a string, $v = \sqrt{T_s/\mu}$. Increasing the tension or decreasing the linear density (lighter string) will increase the wave speed.

Stop to Think 20.3: d. The wavelength—the distance between two crests—is seen to be 10 m. The frequency is $f = v/\lambda = (50 \text{ m/s})/(10 \text{ m}) = 5$ Hz.

Stop to Think 20.4: e. A crest and an adjacent trough are separated by $\lambda/2$. This is a phase difference of π.

Stop to Think 20.5: $n_3 > n_1 > n_2$. $\lambda = \lambda_{vac}/n$, so a shorter wavelength corresponds to a larger index of refraction.

Stop to Think 20.6: c. Zack hears a higher frequency as he and the source approach. Amy is moving with the source, so $f_{Amy} = f_0$.

21 Superposition

► **Looking Ahead**

The goal of Chapter 21 is to understand and use the idea of superposition. In this chapter you will learn to:

- Apply the principle of superposition.
- Understand how standing waves are generated.
- Calculate the allowed wavelengths and frequencies of standing waves.
- Understand how waves cause constructive and destructive interference.
- Calculate the beat frequency between two nearly equal frequencies.

◄ **Looking Back**

The material in this chapter depends on many properties of traveling waves that were introduced in Chapter 20. Please review:

- Sections 20.2–20.4 The fundamental properties of traveling waves.
- Section 20.5 Sound waves and light waves.

This piece of glass has one color if you look through it, a different color if you look at light reflected from it. It is called *dichroic glass,* meaning two-color glass, and it is used both for jewelry and in the electro-optics industry. Ordinary colored glass, such as a green bottle, reflects and transmits the same color of light. Why is dichroic glass different? You'll learn in this chapter that the interesting optical properties of dichroic glass are due to the combination of *two* traveling waves.

The combination of two or more waves is called a *superposition* of waves. In this chapter we will explore how waves are superimposed and learn that superposition is important to applications ranging from musical instruments to lasers. This chapter also lays the groundwork for our study of light-wave optics in Chapter 22.

21.1 The Principle of Superposition

Figure 21.1a shows two baseball players, Alan and Bill, at batting practice. Unfortunately, someone has turned the pitching machines so that pitching machine A throws baseballs toward Bill while machine B throws toward Alan. If two baseballs are launched at the same time, and with the same speed, they collide at the crossing point and bounce away. Two particles cannot occupy the same point of space at the same time.

(a)

(b)

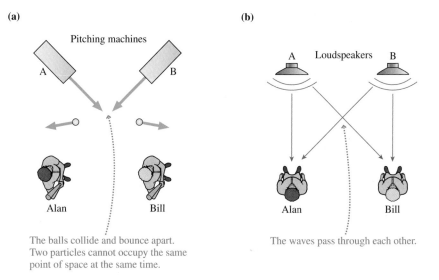

The balls collide and bounce apart. Two particles cannot occupy the same point of space at the same time.

The waves pass through each other.

FIGURE 21.1 Unlike particles, two waves can pass directly through each other.

But waves, unlike particles, can pass directly through each other. In Figure 21.1b Alan and Bill are listening to the stereo system in the locker room after practice. Because both hear the music quite well, without distortion or missing sound, the sound wave that travels from loudspeaker A toward Bill must pass through the wave traveling from loudspeaker B toward Alan. What happens to the medium at a point where two waves are present simultaneously?

If wave 1 displaces a particle in the medium by D_1 and wave 2 *simultaneously* displaces it by D_2, the net displacement of the particle is simply $D_1 + D_2$. This is a very important idea because it tells us how to combine waves. It is known as the *principle of superposition.*

Principle of superposition When two or more waves are *simultaneously* present at a single point in space, the displacement of the medium at that point is the sum of the displacements due to each individual wave.

When different objects are laid on top of each other, they are said to be *superimposed.* But through some quirk in the English language, the result of superimposing objects is called a *superposition,* without the syllable "im" in the middle. When one wave is "placed" on top of another wave, we have a superposition of waves, hence the name of the principle.

Mathematically, the net displacement of a particle in the medium is

$$D_{net} = D_1 + D_2 + \cdots = \sum_i D_i \qquad (21.1)$$

where D_i is the displacement that would be caused by wave i alone. We will make the simplifying assumption that the displacements of the individual waves are along the same line so that we can add displacements as scalars rather than vectors.

To use the principle of superposition you must know the displacement that each wave would cause if it traveled through the medium alone. Then you go through the medium *point by point* and add the displacements due to each wave *at that point* to find the net displacement at that point. The outcome will be different at each and every point in the medium because the displacements are different at each point.

To illustrate, Figure 21.2 shows five snapshot graphs taken 1 s apart of two waves traveling at the same speed (1 m/s) in opposite directions along a string.

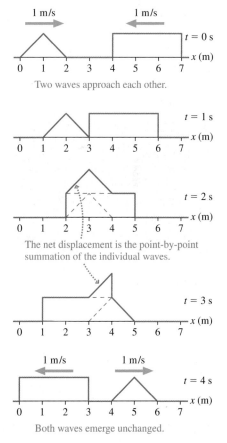

Two waves approach each other.

The net displacement is the point-by-point summation of the individual waves.

Both waves emerge unchanged.

FIGURE 21.2 The superposition of two waves on a string as they pass through each other.

The principle of superposition comes into play wherever the waves overlap. The displacement of each wave is shown as a dotted line. The solid line is the sum *at each point* of the two displacements at that point. This is the displacement that you would actually observe as the two waves pass through each other.

> **NOTE** ▶ Superposition is the key idea for this chapter and the next, so make sure you understand how the solid line is the point-by-point addition of the displacements due to the individual waves. ◀

STOP TO THINK 21.1 Two pulses on a string approach each other at speeds of 1 m/s. What is the shape of the string at $t = 6$ s?

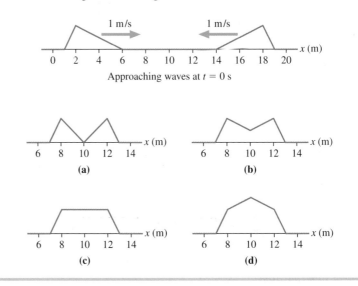

21.2 Standing Waves

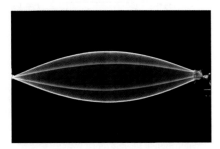

FIGURE 21.3 A vibrating string is an example of a standing wave.

Figure 21.3 is a time-lapse photograph of a *standing wave* on a vibrating string. It's not obvious from the photograph, but a standing wave is actually the superposition of two waves. To understand this, let's begin by thinking about two sinusoidal waves traveling in opposite directions through a medium. For example, suppose you point two loudspeakers at each other or shake both ends of a very long string. Throughout this section, **we will assume that the two waves have the same frequency, the same wavelength, and the same amplitude.** In other words, they're identical waves except that one travels to the right and the other to the left. What happens as these two waves pass through each other?

Figure 21.4a shows nine snapshot graphs, at intervals of $\frac{1}{8}T$, of the two waves moving through each other. The dots identify two of the crests to help you see that the red wave is traveling to the right and the green wave to the left. At *each point,* the net displacement of the medium is found by adding the red displacement and the green displacement. The resulting blue wave is the superposition of the two traveling waves.

Figure 21.4a is rather complicated, so Figure 21.4b shows just the blue superposition of the two waves. This is the wave that you would actually observe in the medium. Interestingly, the blue dot shows that the wave in Figure 21.4b is moving neither right nor left. This is a wave, but it is not a traveling wave. The wave in Figure 21.4b is called a **standing wave** because the crests and troughs "stand in place" as it oscillates.

(a)

The red wave is traveling to the right.

The green wave is traveling to the left.

(b)

The superposition of the red and green waves is a standing wave.

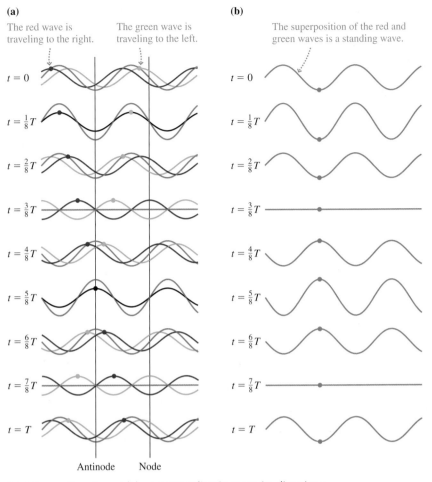

Antinode   Node

FIGURE 21.4 Two sinusoidal waves traveling in opposite directions.

Nodes and Antinodes

Figure 21.5 has collapsed the nine graphs of Figure 21.4b into a single graphical representation of a standing wave. Compare this to the Figure 21.3 photograph of a vibrating string and you can see that the vibrating string is a standing wave. A striking feature of a standing-wave pattern is the existence of points that *never move!* These points, which are spaced $\lambda/2$ apart, are called **nodes.** Halfway between the nodes are the points where the particles in the medium oscillate with maximum displacement. These points of maximum amplitude are called **antinodes,** and you can see that they are also spaced $\lambda/2$ apart.

It seems surprising and counterintuitive that some particles in the medium have no motion at all. To understand how this happens, look carefully at the two traveling waves in Figure 21.4a. You will see that the nodes occur at points where at *every instant* of time the displacements of the two traveling waves have equal magnitudes but *opposite signs*. Thus the superposition of the displacements at these points is always zero. The antinodes correspond to points where the two displacements have equal magnitudes and the *same sign* at all times. The superposition at these points gives a displacement twice that of each individual wave.

Two waves 1 and 2 are said to be *in phase* at a point where D_1 is *always* equal to D_2. The superposition at that point yields a wave whose amplitude is twice that of the individual waves. This is called a point of *constructive interference*. The antinodes of a standing wave are points of constructive interference between the two traveling waves.

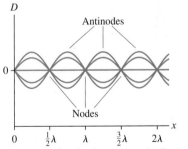

The nodes and antinodes are spaced $\lambda/2$ apart.

FIGURE 21.5 Standing waves are often represented as they would be seen in a time-lapse photograph.

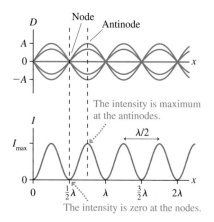

FIGURE 21.6 The intensity of a standing wave is maximum at the antinodes, zero at the nodes.

In contrast, two waves are said to be *out of phase* at points where D_1 is *always* equal to $-D_2$. Their superposition gives a wave with zero amplitude—no wave at all! This is a point of *destructive interference*. The nodes of a standing wave are points of destructive interference. We will defer the main discussion of constructive and destructive interference until later in this chapter, but you'll then recognize that you're seeing constructive and destructive interference at the antinodes and nodes of a standing wave.

In Chapter 20 you learned that the *intensity* of a wave is proportional to the square of the amplitude: $I = CA^2$. You can see in Figure 21.6 that the points of maximum intensity occur where the standing wave oscillates with the largest amplitude (i.e., the antinodes) and that the intensity is zero at the nodes. If this is a sound wave, the loudness periodically varies from zero (no sound) to a maximum and back to zero. The key idea is that **the intensity is maximum at points of constructive interference and zero (if the waves have equal amplitudes) at points of destructive interference.**

The Mathematics of Standing Waves

A sinusoidal wave traveling to the right along the x-axis with angular frequency $\omega = 2\pi f$, wave number $k = 2\pi/\lambda$, and amplitude a is

$$D_R = a\sin(kx - \omega t) \tag{21.2}$$

An equivalent wave traveling to the left is

$$D_L = a\sin(kx + \omega t) \tag{21.3}$$

We previously used the symbol A for the wave amplitude, but here we will use a lower case a to represent the amplitude of each individual wave and reserve A for the amplitude of the net wave.

According to the principle of superposition, the net displacement of the medium when both waves are present is the sum of D_R and D_L:

$$D(x, t) = D_R + D_L = a\sin(kx - \omega t) + a\sin(kx + \omega t) \tag{21.4}$$

We can simplify Equation 21.4 by using the trigonometric identity

$$\sin(\alpha \pm \beta) = \sin\alpha\cos\beta \pm \cos\alpha\sin\beta$$

Doing so gives

$$\begin{aligned} D(x, t) &= a(\sin kx \cos\omega t - \cos kx \sin\omega t) + a(\sin kx \cos\omega t + \cos kx \sin\omega t) \\ &= (2a\sin kx)\cos\omega t \end{aligned} \tag{21.5}$$

It is useful to write Equation 21.5 as

$$D(x, t) = A(x)\cos\omega t \tag{21.6}$$

where the **amplitude function** $A(x)$ is defined as

$$A(x) = 2a\sin kx \tag{21.7}$$

The amplitude reaches a maximum value $A_{max} = 2a$ at points where $\sin kx = 1$.

The displacement $D(x, t)$ given by Equation 21.6 is neither a function of $x - vt$ nor a function of $x + vt$, hence it is *not* a traveling wave. Instead, the $\cos\omega t$ term in Equation 21.6 describes a medium in which each point oscillates in simple harmonic motion with frequency $f = \omega/2\pi$. The function $A(x) = 2a\sin kx$ determines the amplitude of the oscillation for a particle at position x.

Figure 21.7 graphs Equation 21.6 at several different instants of time. Notice that the graphs are identical to those of Figure 21.5, showing us that Equation 21.6 is the mathematical description of a standing wave. You can see that the amplitude of oscillation, given by $A(x)$, varies from point to point in the medium.

The nodes of the standing wave are the points at which the amplitude is zero. They are located at positions x for which

$$A(x) = 2a \sin kx = 0 \qquad (21.8)$$

Equation 21.8 is true if

$$kx_m = \frac{2\pi x_m}{\lambda} = m\pi \qquad m = 0, 1, 2, 3, \ldots \qquad (21.9)$$

Thus the position x_m of the mth node is

$$x_m = m\frac{\lambda}{2} \qquad m = 0, 1, 2, 3, \ldots \qquad (21.10)$$

where m is an integer. You can see that the spacing between two adjacent nodes is $\lambda/2$, in agreement with Figure 21.6. The nodes are *not* spaced by λ, as you might have expected.

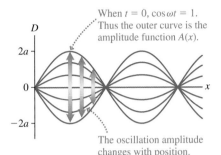

FIGURE 21.7 The net displacement due to two counter-propagating sinusoidal waves.

EXAMPLE 21.1 Node spacing on a string
A very long string has a linear density of 5.0 g/m and is stretched with a tension of 8.0 N. 100 Hz waves with amplitudes of 2.0 mm are generated at the ends of the string.

a. What is the node spacing along the resulting standing wave?
b. What is the maximum displacement of the string?

MODEL Two counter-propagating waves of equal frequency create a standing wave.

VISUALIZE The standing wave will look like Figure 21.5.

SOLVE

a. The speed of the waves on the string is

$$v = \sqrt{\frac{T_s}{\mu}} = \sqrt{\frac{8.0 \text{ N}}{0.0050 \text{ kg/m}}} = 40 \text{ m/s}$$

and thus the wavelength is

$$\lambda = \frac{v}{f} = \frac{40 \text{ m/s}}{100 \text{ Hz}} = 0.40 \text{ m} = 40 \text{ cm}$$

Consequently, the spacing between adjacent nodes is $\lambda/2 = 20$ cm.

b. The maximum displacement, at the antinodes where $\sin kx = 1$, is

$$A_{\text{max}} = 2a = 4.0 \text{ mm}$$

21.3 Transverse Standing Waves

Wiggling both ends of a very long string is not a practical way to generate standing waves. Instead, as in the photograph in Figure 21.3, standing waves are usually seen on a string that is fixed at both ends. To understand why this condition causes standing waves, we need to examine what happens when a traveling wave encounters a discontinuity or a boundary.

Figure 21.8a on the next page shows a *discontinuity* between a string with a larger linear density and one with a smaller linear density. The tension is the same in both strings, so the wave speed is slower on the left, faster on the right. Whenever a wave encounters a discontinuity, some of the wave's energy is *transmitted* forward and some is *reflected*. Here the larger fraction of the energy is transmitted, but you can see that neither the transmitted not the reflected pulse has the same amplitude as the initial wave pulse.

Light waves exhibit an analogous behavior when they encounter a piece of glass. Most of the light wave's energy is transmitted through the glass, which is

Activ Physics   10.4, 10.6

(a)

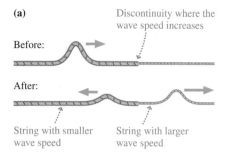

Discontinuity where the wave speed increases

Before:

After:

String with smaller wave speed

String with larger wave speed

(b)

Discontinuity where the wave speed decreases

Before:

After:

The reflected pulse is inverted.

(c)

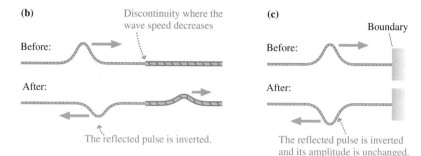

Boundary

Before:

After:

The reflected pulse is inverted and its amplitude is unchanged.

FIGURE 21.8 A wave reflects when it encounters a discontinuity or a boundary.

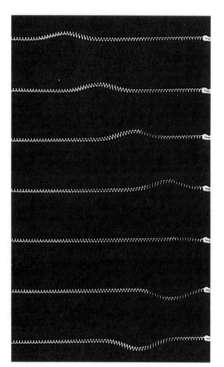

FIGURE 21.9 A strobe photo of a pulse traveling along a rope-like spring.

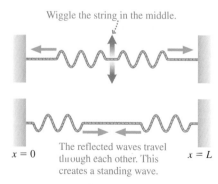

Wiggle the string in the middle.

$x = 0$

The reflected waves travel through each other. This creates a standing wave.

$x = L$

FIGURE 21.10 Reflections at the two boundaries cause a standing wave on the string.

why glass is transparent, but a small amount of energy is reflected. That is how you see your reflection dimly in a storefront window.

In Figure 21.8b, an incident wave encounters a discontinuity at which the wave speed decreases. Once again, some of the wave's energy is transmitted and some is reflected. But notice something interesting. When a wave encounters a discontinuity at which the speed decreases, the reflected pulse is *inverted*. A displacement D on the incident wave becomes displacement $-D$ on the reflected wave. Because $\sin(\phi + \pi) = -\sin\phi$, we say that the reflected wave has a *phase change of π upon reflection*. This aspect of reflection will be important later in the chapter when we look at the interference of light waves.

The wave in Figure 21.8c reflects from a *boundary*. You can think of this as Figure 21.8b in the limit that the string on the right becomes infinitely massive. Thus the reflection in Figure 21.8c looks like that of Figure 21.8b with one exception: Because there is no transmitted wave, *all* the wave's energy is reflected. Hence **the amplitude of a wave reflected from a boundary is unchanged.** Figure 21.9 is a sequence of strobe photos in which you see a pulse on a rope-like spring reflecting from a boundary at the right of the photo. The reflected pulse is inverted but otherwise unchanged.

Standing Waves on a String

Figure 21.10 shows a string of length L that is tied at $x = 0$ and $x = L$. This string has *two* boundaries where reflections can occur. If you wiggle the string in the middle, sinusoidal waves travel outward in both directions and soon reach the boundaries. Because the speed of a reflected wave does not change, **the wavelength and frequency of a reflected sinusoidal wave are unchanged.** Consequently, reflections at the ends of the string cause two waves of *equal amplitude and wavelength* to travel in opposite directions along the string. As we've just seen, these are the conditions that cause a standing wave!

To connect the mathematical analysis of standing waves in Section 21.2 with the physical reality of a string tied down at the ends, we need to impose *boundary conditions*. A **boundary condition** is a mathematical statement of any constraint that *must* be obeyed at the boundary or edge of a medium. Because the string is tied down at the ends, the displacements at $x = 0$ and $x = L$ must be zero at all times. Thus the standing-wave boundary conditions are $D(x = 0, t) = 0$ and $D(x = L, t) = 0$. Stated another way, we require nodes at both ends of the string.

We found that the displacement of a standing wave is $D(x, t) = (2a \sin kx)\cos \omega t$. You can see that this equation already satisfies the boundary condition $D(x = 0, t) = 0$. That is, the origin has already been located at a node. The second boundary condition, at $x = L$, requires $D(x = L, t) = 0$. This condition will be met at all times if

$$2a \sin kL = 0 \qquad \text{(boundary condition at } x = L) \qquad (21.11)$$

Equation 21.11 will be true if $\sin kL = 0$, which in turn requires

$$kL = \frac{2\pi L}{\lambda} = m\pi \qquad m = 1, 2, 3, 4, \ldots \qquad (21.12)$$

kL must be a multiple of $m\pi$, but $m = 0$ is excluded because L can't be zero.

For a string of fixed length L, the only quantity in Equation 21.12 that can vary is λ. That is, the boundary condition can be satisfied only if the wavelength has one of the values

$$\lambda_m = \frac{2L}{m} \qquad m = 1, 2, 3, 4, \ldots \qquad (21.13)$$

In other words, **a standing wave can exist on the string *only* if its wavelength is one of the values given by Equation 21.13.** The mth possible wavelength $\lambda_m = 2L/m$ is just the right size so that its mth node is located at the end of the string (at $x = L$).

NOTE ▶ Other wavelengths, which would be perfectly acceptable wavelengths for a traveling wave, cannot exist as a *standing* wave of length L because they cannot meet the boundary conditions requiring a node at each end of the string. ◀

If standing waves are possible only for certain wavelengths, then only a few specific oscillation frequencies are allowed. Because $\lambda f = v$ for a sinusoidal wave, the oscillation frequency corresponding to wavelength λ_m is

$$f_m = \frac{v}{\lambda_m} = \frac{v}{2L/m} = m\frac{v}{2L} \qquad m = 1, 2, 3, 4, \ldots \qquad (21.14)$$

The lowest allowed frequency

$$f_1 = \frac{v}{2L} \qquad \text{(fundamental frequency)} \qquad (21.15)$$

which corresponds to wavelength $\lambda_1 = 2L$, is called the **fundamental frequency** of the string. The allowed frequencies can be written in terms of the fundamental frequency as

$$f_m = mf_1 \qquad m = 1, 2, 3, 4, \ldots \qquad (21.16)$$

The allowed standing-wave frequencies are all integer multiples of the fundamental frequency. The higher-frequency standing waves are called **harmonics,** with the $m = 2$ wave at frequency f_2 called the *second harmonic,* the $m = 3$ wave called the *third harmonic,* and so on.

Figure 21.11 graphs the first four possible standing waves on a string of fixed length L. These possible standing waves are called the **normal modes** of the string. Each mode, numbered by the integer m, has a unique wavelength and frequency. Keep in mind that these drawings simply show the *envelope,* or outer edge, of the oscillations. The string is continuously oscillating at all positions between these edges, as we showed in more detail in Figure 21.5.

There are several things to note about the normal modes of a string.

1. m is the number of *antinodes* on the standing wave, not the number of nodes. You can tell a string's mode of oscillation by counting the number of antinodes.
2. The *fundamental mode,* with $m = 1$, has $\lambda_1 = 2L$, not $\lambda_1 = L$. Only half of a wavelength is contained between the boundaries, a direct consequence of the fact that the spacing between nodes is $\lambda/2$.

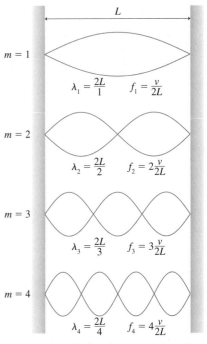

FIGURE 21.11 The first four normal modes for standing waves on a string of length L.

FIGURE 21.12 Time-exposure photograph of the $m = 4$ standing-wave mode on a stretched string.

3. The frequencies of the normal modes form an arithmetic series: $f_1, 2f_1, 3f_1, 4f_1, \ldots$. The fundamental frequency f_1 can be found as the *difference* between the frequencies of any two adjacent modes. That is, $f_1 = \Delta f = f_{m+1} - f_m$.

Figure 21.12 is a time-exposure photograph of the $m = 4$ standing wave on a string. The nodes and antinodes are quite distinct. The string vibrates four times faster for the $m = 4$ mode than for the fundamental $m = 1$ mode.

EXAMPLE 21.2   A standing wave on a string

A 2.50-m-long string vibrates as a 100 Hz standing wave with nodes 1.00 m and 1.50 m from one end of the string and at no points in between these two. Which harmonic is this, and what is the string's fundamental frequency?

MODEL   The nodes of a standing wave are spaced $\lambda/2$ apart.

VISUALIZE   The standing wave looks like Figure 21.5.

SOLVE   If there are no nodes between the two at 1.0 m and 1.5 m, then the node spacing is $\lambda/2 = 0.50$ m. The number of 0.50-m-wide segments that fit into a 2.50 m length is five, so this is the $m = 5$ mode and 100 Hz is the fifth harmonic. The harmonic frequencies are $f_m = mf_1$, hence the fundamental frequency is

$$f_1 = \frac{f_5}{5} = \frac{100 \text{ Hz}}{5} = 20 \text{ Hz}$$

STOP TO THINK 21.2   A standing wave on a string vibrates as shown at the right. Suppose the tension is quadrupled while the frequency and the length of the string are held constant. Which standing-wave pattern is produced?

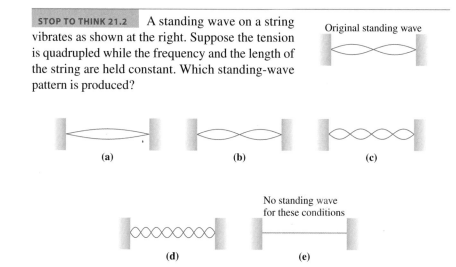

Original standing wave

(a)   (b)   (c)

No standing wave for these conditions

(d)   (e)

Standing Electromagnetic Waves

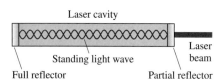

Laser cavity

Standing light wave

Full reflector

Laser beam

Partial reflector

FIGURE 21.13 A laser contains a standing light wave between two parallel mirrors.

A vibrating string is only one example of a transverse standing wave. Another transverse wave is an electromagnetic wave. Standing electromagnetic waves can be established between two parallel mirrors that reflect light back and forth. The mirrors are boundaries, analogous to the boundaries at the ends of a string. In fact, this is exactly how a laser operates. The two facing mirrors in Figure 21.13 form what is called a *laser cavity.*

Because the mirrors act exactly like the points to which a string is tied, the boundary condition is that the light wave must have a node at the surface of each mirror. (To allow some of the light to escape the laser cavity and form the *laser beam,* one of the mirrors is only partially reflective. This doesn't affect the boundary condition.)

Because the boundary conditions are the same, Equations 21.13 and 21.14 for λ_m and f_m apply to a laser just as they do to a vibrating string. The primary difference is the size of the wavelength. A typical laser cavity has a length $L \approx 30$ cm, and visible light has a wavelength $\lambda \approx 600$ nm. The standing light wave in a laser cavity has a mode number m that is approximately

$$m = \frac{2L}{\lambda} \approx \frac{2 \times 0.30 \text{ m}}{6.00 \times 10^{-7} \text{ m}} = 1{,}000{,}000$$

In other words, the standing light wave inside a laser cavity has approximately one million antinodes! This is a consequence of the very small wavelength of light.

EXAMPLE 21.3 The standing light wave inside a laser
Helium-neon lasers emit the red laser light commonly used in classroom demonstrations and supermarket checkout scanners. A helium-neon laser operates at a wavelength of precisely 632.9924 nm when the spacing between the mirrors is 310.372 mm.

a. In which mode does this laser operate?
b. What is the next longest wavelength that could form a standing wave in this laser cavity?

MODEL The light wave forms a standing wave between the two mirrors.

VISUALIZE The standing wave looks like Figure 21.13.

SOLVE

a. We can use $\lambda_m = 2L/m$ to find that m (the mode) is

$$m = \frac{2L}{\lambda_m} = \frac{2 \times 0.310372 \text{ m}}{6.329924 \times 10^{-7} \text{ m}} = 980{,}650$$

There are 980,650 antinodes in the standing light wave between the mirrors.

b. The next longest wavelength that can fit in this laser cavity will have one less node. It will be the $m = 980{,}649$ mode and its wavelength will be

$$\lambda = \frac{2L}{m} = \frac{2(0.310372 \text{ m})}{980{,}649} = 632.9930 \text{ nm}$$

ASSESS The mode number is very large, as expected. The wavelength increases by a mere 0.0006 nm when the mode number is decreased by 1.

Microwave radiation, which has a wavelength of a few centimeters, can also set up standing waves between reflective surfaces. This is not always good. If the microwaves in a microwave oven form a standing wave, there are nodes where the electromagnetic field intensity is always zero. These nodes cause cold spots where the food does not heat. Although designers of microwave ovens try to prevent standing waves, ovens usually do have cold spots spaced $\lambda/2$ apart at nodes in the microwave field. A turntable in a microwave oven keeps the food moving so that no part of your dinner remains at a node.

EXAMPLE 21.4 Cold spots in a microwave oven
Cold spots in a microwave oven are found to be 6.0 cm apart. What is the frequency of the microwaves?

MODEL A standing wave is created by microwaves reflecting from the walls.

SOLVE The cold spots are nodes in the microwave standing wave. Nodes are spaced $\lambda/2$ apart, so the wavelength of the microwave radiation must be $\lambda = 12$ cm $= 0.12$ m. The speed of microwaves is the speed of light, $v = c$, so the frequency is

$$f = \frac{c}{\lambda} = \frac{3.00 \times 10^8 \text{ m/s}}{0.12 \text{ m}} = 2.5 \times 10^9 \text{ Hz} = 2.5 \text{ GHz}$$

The standing wave must have a node at a closed end.

Air molecules undergo longitudinal oscillations.

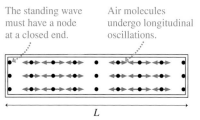

L

Physical representation of the m = 2 mode

Graphical representation

FIGURE 21.14 The $m = 2$ longitudinal standing wave inside a closed column of air.

21.4 Standing Sound Waves and Musical Acoustics

A long, narrow column of air, such as the air in a tube or a pipe, can support a *longitudinal* standing sound wave. As a sound wave travels through a tube, the air oscillates *parallel* to the tube. If the tube is closed at the end, with a cap, the air at the closed end cannot oscillate. Consequently, **the closed end of a column of air must be a node.**

Figure 21.14 shows the $m = 2$ standing wave inside a column of air closed at both ends. We call this a *closed-closed tube.* You can see that the displacement graph looks exactly like the $m = 2$ standing-wave mode on a string. Because there must be a node at both ends, the possible wavelengths and frequencies of a closed air column are the same as for a string of the same length.

> **EXAMPLE 21.5 Singing in the shower**
> A shower stall is 2.45 m (8 ft) tall. For what frequencies less than 500 Hz are there standing sound waves in the shower stall?
>
> **MODEL** The shower stall, at least to a first approximation, is a column of air 2.45 m long. It is closed at the ends by the ceiling and floor. Assume a room-temperature speed of sound.
>
> **VISUALIZE** A standing sound wave will have nodes at the ceiling and the floor. The $m = 2$ mode will look like Figure 21.14 rotated 90°.
>
> **SOLVE** The fundamental frequency for a standing sound wave in this air column is
>
> $$f_1 = \frac{v}{2L} = \frac{343 \text{ m/s}}{2(2.45 \text{ m})} = 70 \text{ Hz}$$
>
> The possible standing-wave frequencies are integer multiples of the fundamental frequency. These are 70 Hz, 140 Hz, 210 Hz, 280 Hz, 350 Hz, 420 Hz, and 490 Hz.
>
> **ASSESS** The many possible standing waves in a shower cause the sound to *resonate,* which helps explain why some people like to sing in the shower. Our approximation of the shower stall as a one-dimensional tube is actually a bit too simplistic. A three-dimensional analysis would find additional modes, making the "sound spectrum" even richer.

Air columns closed at both ends are of limited interest unless, as in Example 21.5, you are inside the column. Columns of air that *emit* sound are open at one or both ends. Many musical instruments fit this description. For example, a flute is a tube of air that is open at both ends. The flutist blows across one end to create a standing wave inside the tube, and a note of that frequency is emitted from both ends of the flute. (The blown end of a flute is open on the side, rather than across the tube. That is necessary for practical reasons of how flutes are played, but from a physics perspective this is the "end" of the tube because it opens the tube to the atmospheric pressure of the surrounding air.) A trumpet, however, is open at the bell end but is *closed* by the player's lips at the other end.

You saw earlier that a wave is partially transmitted and partially reflected at a discontinuity. When a sound wave traveling through a tube of air reaches an open end, some of the wave's energy is transmitted out of the tube to become the sound that you hear and some portion of the wave is reflected back into the tube. These reflections, analogous to the reflection of a string wave from a boundary, allow standing sound waves to exist in a tube of air that is open at one or both ends.

Not surprisingly, the *boundary condition* at the open end of a column of air is not the same as the boundary condition at a closed end. We've seen that the closed

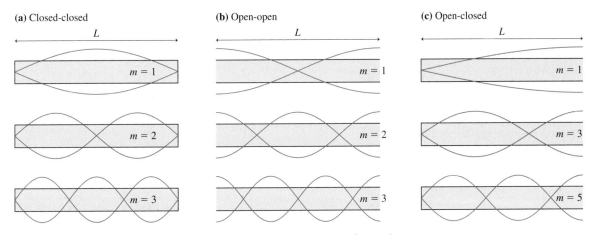

(a) Closed-closed

(b) Open-open

(c) Open-closed

FIGURE 21.15 The first three standing sound wave modes in columns of air with different boundary conditions.

end must be a node because the air has no room to vibrate. By contrast, **an open end of an air column is required to be an antinode.** (A careful analysis shows that the antinode is actually just outside the open end, but for our purposes we'll assume the antinode is exactly at the open end.)

Figure 21.15 shows displacement graphs of the first three standing-wave modes of a tube closed at both ends (a *closed-closed tube*), a tube open at both ends (an *open-open tube*), and a tube open at one end but closed at the other (an *open-closed tube*), all with the same length L. The standing wave in the open-open tube looks like the closed-closed tube except that the positions of the nodes and antinodes are interchanged. In both cases there are m half-wavelength segments between the ends, thus the wavelengths and frequencies of an open-open tube and a closed-closed tube are the same as those of a string tied at both ends:

$$\begin{cases} \lambda_m = \dfrac{2L}{m} \\ f_m = m\dfrac{v}{2L} = mf_1 \end{cases} \quad \begin{matrix} m = 1, 2, 3, 4, \ldots \\ \text{(open-open or closed-closed tube)} \end{matrix} \qquad (21.17)$$

The open-closed tube is different. The fundamental mode has only one-quarter of a wavelength in a tube of length L, hence the $m = 1$ wavelength is $\lambda_1 = 4L$. This is twice the λ_1 wavelength of an open-open or a closed-closed tube. Consequently, **the fundamental frequency of an open-closed tube is half that of an open-open or a closed-closed tube of the same length.** It will be left as a homework problem for you to show that the possible wavelengths and frequencies of an open-closed tube of length L are

$$\begin{cases} \lambda_m = \dfrac{4L}{m} \\ f_m = m\dfrac{v}{4L} = mf_1 \end{cases} \quad \begin{matrix} m = 1, 3, 5, 7, \ldots \\ \text{(open-closed tube)} \end{matrix} \qquad (21.18)$$

Notice that m in Equation 21.18 takes on only *odd* values.

NOTE ▶ Because sound is a longitudinal wave, the graphs of Figure 21.15 are *not* "pictures" of the wave as they are for a string wave. The graphs show the displacement Δx, *parallel* to the axis, versus position x. The tube itself is shown merely to indicate the location of the open and closed ends, but the diameter of the tube is *not* related to the amplitude of the displacement. ◀

EXAMPLE 21.6 The length of an organ pipe

An organ pipe open at both ends sounds its second harmonic at a frequency of 523 Hz. (Musically, this is the note one octave above middle C.) What is the length of the pipe from the sounding hole to the end?

MODEL An organ pipe, similar to a flute, has a *sounding hole* where compressed air is blown across the edge of the pipe. This is one end of an open-open tube, with the other end at the true "end" of the pipe. Assume a room-temperature speed of sound.

SOLVE The second harmonic is the $m = 2$ mode, which for an open-open tube has frequency

$$f_2 = 2\frac{v}{2L}$$

Thus the length of the organ pipe is

$$L = \frac{v}{f_2} = \frac{343 \text{ m/s}}{523 \text{ Hz}} = 0.656 \text{ m} = 65.6 \text{ cm}$$

ASSESS This is a typical length for an organ pipe.

STOP TO THINK 21.3 An open-open tube of air supports standing waves at frequencies of 300 Hz and 400 Hz and at no frequencies between these two. The second harmonic of this tube has frequency

a. 100 Hz.   b. 200 Hz.   c. 400 Hz.   d. 600 Hz.   e. 800 Hz.

Musical Instruments

10.5   Activ Physics

An important application of standing waves is to musical instruments. Think about stringed musical instruments, such as the guitar, the piano, and the violin. These instruments all have strings that are fixed at the ends and tightened to create tension. A disturbance is generated on the string by plucking, striking, or bowing it. Regardless of how it is generated, the disturbance creates a standing wave on the string.

The fundamental frequency of a vibrating string is

$$f_1 = \frac{v}{2L} = \frac{1}{2L}\sqrt{\frac{T_s}{\mu}}$$

where T_s is the tension in the string and μ is its linear density. The fundamental frequency is the musical note you hear when the string is sounded. Increasing the tension in the string raises the fundamental frequency, a fact known to anyone who has ever tuned a stringed instrument.

NOTE ▶ v is the wave speed *on the string,* not the speed of sound in air. ◀

For instruments like the guitar or the violin, the strings are all the same length and under approximately the same tension. Were that not the case, the neck of the instrument would tend to twist toward the side of higher tension. The strings have different frequencies because they differ in linear density. The lower-pitched strings are "fat" while the higher-pitched strings are "skinny." This difference changes the frequency by changing the wave speed. *Small* adjustments are then made in the tension to bring each string to the exact desired frequency. Once the instrument is tuned, you play it by using your finger tips to alter the effective length of the string. As you shorten the string's length, the frequency and pitch go up.

A piano covers a much wider range of frequencies than a guitar or violin. This range cannot be produced by changing only the linear densities of the strings. The high end would have strings too thin to use without breaking, and the low end would have solid rods rather than flexible wires! So a piano is tuned through a combination of changing the linear density *and* the length of the strings. The bass note strings are not only fatter, they are also longer.

With a wind instrument, blowing into the mouthpiece creates a standing sound wave inside a tube of air. The player changes the notes by using her fingers to cover holes or open valves, changing the length of the tube and thus its frequency.

The strings on a harp vibrate as standing waves. Their frequencies determine the notes that you hear.

As we noted about the flute, the fact that the holes are on the side makes very little difference. The first open hole becomes an antinode because the air is free to oscillate in and out of the opening.

According to Equations 21.17 and 21.18, a wind instrument's frequency depends on the speed of sound *inside* the instrument. But the speed of sound depends on the temperature of the air. When a wind player first blows into the instrument, the air inside starts to rise in temperature. This increases the sound speed, which in turn raises the instrument's frequency for each note until the air temperature reaches a steady state. Consequently, wind players must "warm up" before tuning their instrument. For strings, the speed appearing in Equation 21.17 is the wave speed on the string as determined by the tension, not the sound speed in air.

Many wind instruments have a "buzzer" at one end of the tube, such as a vibrating reed on a saxophone or clarinet or vibrating lips on a trumpet or trombone. Buzzers like these generate a continuous range of frequencies rather than single notes, which is why they sound like a "squawk" if you play on just the mouthpiece without the rest of the instrument. When connected to the body of the instrument, most of those frequencies cause no response of the air molecules. But the frequency from the buzzer that matches the fundamental frequency of the instrument causes the build-up of a large-amplitude response at just that frequency—a standing-wave resonance. This is the energy input that generates and sustains the musical note.

EXAMPLE 21.7 **The notes on a clarinet**
A clarinet is 66 cm long. The speed of sound in warm air is 350 m/s. What are the frequencies of the lowest note on a clarinet and of the next highest harmonic?

MODEL A clarinet is an open-closed tube because the player's lips and the reed seal the tube at the upper end.

SOLVE The lowest frequency is the fundamental frequency. For an open-closed tube, the fundamental frequency is

$$f_1 = \frac{v}{4L} = \frac{350 \text{ m/s}}{4(0.66 \text{ m})} = 133 \text{ Hz}$$

An open-closed tube has only *odd* harmonics. The next highest harmonic is $m = 3$, with frequency $f_3 = 3f_1 = 399$ Hz.

Except in unusual situations, a vibrating string plays only the musical note corresponding to the fundamental frequency f_1. Thus stringed instruments must use several strings in order to obtain a reasonable range of notes. By contrast, wind instruments can sound at the second or third harmonic of the tube of air (f_2 or f_3). These higher frequencies are sounded by *overblowing* (flutes, brass instruments) or with special *register keys* that open small holes in the side of the instrument to encourage the formation of an antinode at that point (clarinets, saxophones). The controlled use of these higher harmonics gives wind instruments a wide range of notes.

21.5 Interference in One Dimension

One of the most basic characteristics of waves is the ability of two waves to combine into a single wave whose displacement is given by the principle of superposition. This combination, or superposition, of waves is often called **interference.** A standing wave is the interference pattern produced when two waves of equal frequency travel in opposite directions. In this section we will look at the interference of two waves traveling in the *same* direction. Then, after the fundamental ideas are established, we will expand our discussion to interference in two and three dimensions.

Figure 21.16a shows two light waves impinging on a partially silvered mirror. Such a mirror partially transmits and partially reflects each wave, causing two *overlapped* light waves to travel along the *x*-axis to the right of the mirror. Or consider the two loudspeakers in Figure 21.16b. The sound wave from loudspeaker 2 passes just to the side of loudspeaker 1, hence two overlapped sound waves travel to the right along the *x*-axis. We want to find out what happens when two overlapped waves travel in the same direction along the same axis.

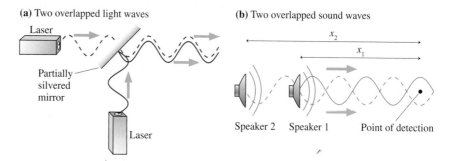

(a) Two overlapped light waves

(b) Two overlapped sound waves

FIGURE 21.16 Two overlapped waves travel along the *x*-axis.

Figure 21.16b shows a point on the *x*-axis where the overlapped waves are detected, either by your ear or by a microphone. This point is distance x_1 from loudspeaker 1 and distance x_2 from loudspeaker 2. (We will use loudspeakers and sound waves for most of our examples, but our analysis will be valid for any wave.) What is the amplitude of the combined waves at this point?

Throughout this section, **we will assume that the waves are sinusoidal, have the same frequency and amplitude, and travel to the right along the *x*-axis.** Thus we can write the displacements of the two waves as

$$D_1(x_1, t) = a\sin(kx_1 - \omega t + \phi_{10}) = a\sin\phi_1$$
$$D_2(x_2, t) = a\sin(kx_2 - \omega t + \phi_{20}) = a\sin\phi_2 \qquad (21.19)$$

where ϕ_1 and ϕ_2 are the *phases* of the waves. Both waves have the same amplitude a, the same wave number $k = 2\pi/\lambda$, and the same angular frequency $\omega = 2\pi f$.

The phase constants ϕ_{10} and ϕ_{20} are characteristics *of the sources,* not the medium. Figure 21.17 shows snapshot graphs at $t = 0$ of waves emitted by three sources with phase constants $\phi_0 = 0$ rad, $\phi_0 = \pi/2$ rad, and $\phi_0 = \pi$ rad. You can see that **the phase constant tells us what the source is doing at $t = 0$.** For example, a loudspeaker at its center position and moving backward at $t = 0$ has $\phi_0 = 0$ rad. Looking back at Figure 21.16b, you can see that loudspeaker 1 has phase constant $\phi_{10} = 0$ rad and loudspeaker 2 has phase constant $\phi_{20} = \pi$ rad.

NOTE ▶ We will often consider *identical sources,* by which we mean that $\phi_{20} = \phi_{10}$. ◀

Let's examine overlapped waves graphically before diving into the mathematics. Figure 21.18 shows two important situations. In part a, the crests of the two waves are aligned as they travel along the *x*-axis. In part b, the crests of one wave align with the troughs of the other wave. The graphs and the wave fronts are slightly displaced from each other so that you can see what each wave is doing, but the *physical situation* is one in which the waves are traveling *on top of* each other. Recall, from Chapter 20, that the wave fronts shown in the middle panel locate the crests of the waves.

The two waves of Figure 21.18a have the same displacement at every point: $D_1(x) = D_2(x)$. Consequently, they must have the same phase. That is, $\phi_1 = \phi_2$

(a) Snapshot graph at $t = 0$ for $\phi_0 = 0$ rad

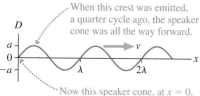

When this crest was emitted, a quarter cycle ago, the speaker cone was all the way forward.

Now this speaker cone, at $x = 0$, is centered and moving backward.

(b) Snapshot graph at $t = 0$ for $\phi_0 = \pi/2$ rad

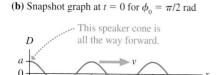

This speaker cone is all the way forward.

(c) Snapshot graph at $t = 0$ for $\phi_0 = \pi$ rad

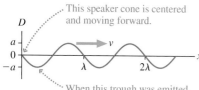

This speaker cone is centered and moving forward.

When this trough was emitted, a quarter cycle ago, the speaker cone was all the way back.

FIGURE 21.17 Waves from three sources having phase constants $\phi_0 = 0$ rad, $\phi_0 = \pi/2$ rad, and $\phi_0 = \pi$ rad.

or, more precisely, $\phi_1 = \phi_2 \pm 2\pi m$ where m is an integer. Two waves that are aligned crest-to-crest and trough-to-trough are said to be **in phase.** Waves that are in phase march along "in step" with each other.

When we combine two in-phase waves, using the principle of superposition, the net displacement at each point is twice the displacement of each individual wave. The superposition of two waves to create a wave with an amplitude *larger* than either individual wave is called **constructive interference.** When the waves are exactly in phase, giving $A = 2a$, we have *maximum constructive interference.*

In Figure 21.18b, where the crests of one wave align with the troughs of the other, the waves march along "out of step" with $D_1(x) = -D_2(x)$ at every point. Two waves that are aligned crest-to-trough are said to be *180° out of phase* or, more generally, just **out of phase.** A superposition of two waves that creates a wave with an amplitude smaller than either individual wave is called **destructive interference.** In this case, because $D_1 = -D_2$, the net displacement is *zero* at *every point* along the axis. The combination of two waves that cancel each other to give no wave is called *perfect destructive interference.*

> **NOTE** ▶ Perfect destructive interference occurs only if the two waves have equal amplitudes, as we're assuming. Two waves of unequal amplitudes can interfere destructively if they have equal wavelengths, but the cancellation won't be perfect. ◀

The Phase Difference

To understand interference, we need to focus on the *phases* of the two waves, which are

$$\phi_1 = kx_1 - \omega t + \phi_{10}$$
$$\phi_2 = kx_2 - \omega t + \phi_{20} \tag{21.20}$$

The difference between the two phases ϕ_1 and ϕ_2, called the **phase difference** $\Delta\phi$, is

$$\begin{aligned} \Delta\phi &= \phi_2 - \phi_1 = (kx_2 - \omega t + \phi_{20}) - (kx_1 - \omega t + \phi_{10}) \\ &= k(x_2 - x_1) + (\phi_{20} - \phi_{10}) \\ &= 2\pi\frac{\Delta x}{\lambda} + \Delta\phi_0 \end{aligned} \tag{21.21}$$

You can see that there are two contributions to the phase difference. $\Delta x = x_2 - x_1$, the distance between the two sources, is called **path-length difference.** It is the extra distance traveled by wave 2 on the way to the point where the two waves are combined. $\Delta\phi_0 = \phi_{20} - \phi_{10}$ is the *inherent phase difference* between the sources. Although the individual phases are dependent on time, the time variable drops out of the phase difference.

The condition of being in phase, where crests are aligned with crests and troughs with troughs, is that $\Delta\phi = 0, 2\pi, 4\pi$, or any integer multiple of 2π. Thus the condition for maximum constructive interference is

> Maximum constructive interference:
>
> $$\Delta\phi = 2\pi\frac{\Delta x}{\lambda} + \Delta\phi_0 = 2m\pi \text{ rad} \qquad m = 0, 1, 2, 3, \dots \tag{21.22}$$

For identical sources, which have $\Delta\phi_0 = 0$ rad, maximum constructive interference occurs when $\Delta x = m\lambda$. That is, **two identical sources produce maximum constructive interference when the path-length difference is an integer number of wavelengths.**

(a) Constructive interference

These two waves are in phase. Their crests are aligned.

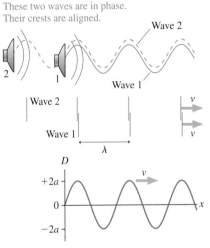

Their superposition produces a wave with amplitude $2a$. This is constructive interference.

(b) Destructive interference

These two waves are out of phase. The crests of one wave are aligned with the troughs of the other.

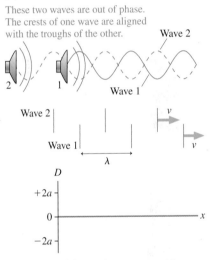

Their superposition produces a wave with zero amplitude. This is destructive interference.

FIGURE 21.18 Constructive and destructive interference of two waves traveling along the x-axis.

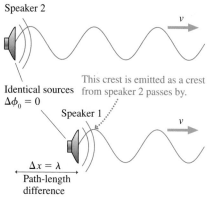

Speaker 2

This crest is emitted as a crest from speaker 2 passes by.

Identical sources
$\Delta\phi_0 = 0$

Speaker 1

$\Delta x = \lambda$
Path-length
difference

The two waves are in phase ($\Delta\phi = 2\pi$ rad) and interfere constructively.

FIGURE 21.19 Two identical sources one wavelength apart.

Figure 21.19 shows two identical sources (i.e., the two loudspeakers are doing the same thing at the same time), so $\Delta\phi_0 = 0$ rad. The path-length difference Δx is the extra distance traveled by the wave from loudspeaker 2 before it combines with loudspeaker 1. In this case, $\Delta x = \lambda$. Because a wave moves forward exactly one wavelength during a time interval of one period, loudspeaker 1 emits a crest exactly as a crest of wave 2 passes by. The two waves are "in step," with $\Delta\phi = 2\pi$ rad, so the two waves interfere constructively to produce a wave of amplitude $2a$.

Perfect destructive interference, where the crests of one wave are aligned with the troughs of the other, occurs when two waves are *out of phase,* meaning that $\Delta\phi = \pi$, 3π, 5π, or any odd multiple of π. Thus the condition for perfect destructive interference is

Perfect destructive interference:

$$\Delta\phi = 2\pi\frac{\Delta x}{\lambda} + \Delta\phi_0 = 2\left(m + \frac{1}{2}\right)\pi \text{ rad} \qquad m = 0, 1, 2, 3, \ldots \tag{21.23}$$

For identical sources, which have $\Delta\phi_0 = 0$ rad, perfect destructive interference occurs when $\Delta x = (m + \frac{1}{2})\lambda$. **That is, two identical sources produce perfect destructive interference when the path-length difference is a half-integer number of wavelengths.**

Two waves can be out of phase because the sources are located at different positions, because the sources themselves are out of phase, or because of a combination of these two. Figure 21.20 illustrates these ideas by showing three different ways in which two waves interfere destructively. Each of these three arrangements creates waves with $\Delta\phi = \pi$ rad.

(a) The sources are out of phase.

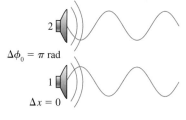

2

$\Delta\phi_0 = \pi$ rad

1

$\Delta x = 0$

(b) Identical sources are separated by half a wavelength.

2

$\Delta\phi_0 = 0$ rad

1

$\Delta x = \frac{1}{2}\lambda$

(c) The sources are both separated and partially out of phase.

2

$\Delta\phi_0 = \frac{\pi}{2}$ rad

1

$\Delta x = \frac{1}{4}\lambda$

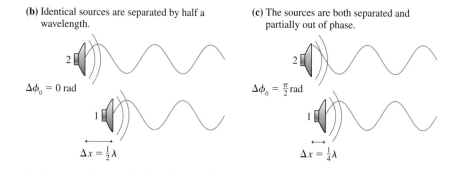

FIGURE 21.20 Destructive interference three ways.

NOTE ▶ Don't confuse the phase difference of the waves ($\Delta\phi$) with the phase difference of the sources ($\Delta\phi_0$). It is $\Delta\phi$, the phase difference of the waves, that governs interference. ◀

EXAMPLE 21.8 **Interference between two sound waves**
You are standing in front of two side-by-side loudspeakers playing sounds of the same frequency. Initially there is almost no sound at all. Then one of the speakers is moved slowly away from you. The sound intensity increases as the separation between the speakers increases, reaching a maximum when the speakers are 0.75 m apart. Then, as the speaker continues to move, the sound starts to decrease. What is the distance between the speakers when the sound intensity is again a minimum?

MODEL The changing sound intensity is due to the interference of two overlapped sound waves.

VISUALIZE Moving one speaker relative to the other, as in Figure 21.20, changes the phase difference between the waves.

SOLVE A minimum sound intensity implies that the two sound waves are interfering destructively. Initially the loudspeakers are side-by-side, so the situation is as shown in Figure 21.20a with $\Delta x = 0$ and $\Delta\phi_0 = \pi$ rad. That is, the speakers themselves are out of phase. Moving one of the speakers does not change $\Delta\phi_0$, but it does change the path-length difference Δx and thus increases the overall phase difference $\Delta\phi$. Constructive interference, causing maximum intensity, is reached when

$$\Delta\phi = 2\pi\frac{\Delta x}{\lambda} + \Delta\phi_0 = 2\pi\frac{\Delta x}{\lambda} + \pi = 2\pi \text{ rad}$$

where we used $m = 1$ because this is the first separation giving constructive interference. The speaker separation at which this occurs is $\Delta x = \lambda/2$. This is the situation shown in Figure 21.21.

Because $\Delta x = 0.75$ m is $\lambda/2$, the sound's wavelength is $\lambda = 1.50$ m. The next point of destructive interference, with $m = 1$, occurs when

$$\Delta\phi = 2\pi\frac{\Delta x}{\lambda} + \Delta\phi_0 = 2\pi\frac{\Delta x}{\lambda} + \pi = 3\pi \text{ rad}$$

Thus the distance between the speakers when the sound intensity is again a minimum is

$$\Delta x = \lambda = 1.50 \text{ m}$$

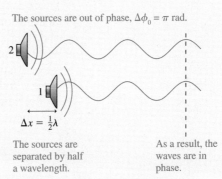

The sources are out of phase, $\Delta\phi_0 = \pi$ rad.

$\Delta x = \frac{1}{2}\lambda$

The sources are separated by half a wavelength.

As a result, the waves are in phase.

FIGURE 21.21 Two out-of-phase sources generate waves that are in phase if the sources are one half-wavelength apart.

ASSESS A separation of λ gives constructive interference for two *identical* speakers ($\Delta\phi_0 = 0$). Here the phase difference of π rad between the speakers (one is pushing forward as the other pulls back) gives destructive interference at this separation.

STOP TO THINK 21.4   Two loudspeakers emit waves with $\lambda = 2.0$ m. Speaker 2 is 1.0 m in front of speaker 1. What, if anything, must be done to cause constructive interference between the two waves?

a. Move speaker 1 forward (to the right) 1.0 m.
b. Move speaker 1 forward (to the right) 0.5 m.
c. Move speaker 1 backward (to the left) 0.5 m.
d. Move speaker 1 backward (to the left) 1.0 m.
e. Nothing. The situation shown already causes constructive interference.
f. Constructive interference is not possible for any placement of the speakers.

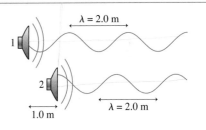

$\lambda = 2.0$ m

$\lambda = 2.0$ m

1.0 m

21.6 The Mathematics of Interference

Let's look more closely at the superposition of two waves . As two waves of equal amplitude and frequency travel together along the x-axis, the net displacement of the medium is

$$D = D_1 + D_2 = a\sin(kx_1 - \omega t + \phi_{10}) + a\sin(kx_2 - \omega t + \phi_{20})$$
$$= a\sin\phi_1 + a\sin\phi_2 \qquad (21.24)$$

where the phases ϕ_1 and ϕ_2 were defined in Equation 21.20.

A useful trigonometric identity is

$$\sin\alpha + \sin\beta = 2\cos\left[\frac{1}{2}(\alpha - \beta)\right]\sin\left[\frac{1}{2}(\alpha + \beta)\right] \qquad (21.25)$$

This identity is certainly not obvious, although it is easily proven by working backward from the right side. We can use this identity to write the net displacement of Equation 21.24 as

$$D = \left[2a\cos\frac{\Delta\phi}{2}\right]\sin(kx_{\text{avg}} - \omega t + (\phi_0)_{\text{avg}}) \qquad (21.26)$$

where $\Delta\phi = \phi_2 - \phi_1$ is the phase difference between the two waves, exactly as in Equation 21.21. $x_{avg} = (x_1 + x_2)/2$ is the average distance to the two sources and $(\phi_0)_{avg} = (\phi_{10} + \phi_{20})/2$ is the average phase constant of the sources.

The sine term shows that the superposition of the two waves is still a traveling wave. An observer would see a sinusoidal wave moving along the x-axis with the *same* wavelength and frequency as the original waves.

But how *big* is this wave compared to the two original waves? They each had amplitude a, but the amplitude of their superposition is

$$A = \left| 2a\cos\left(\frac{\Delta\phi}{2}\right) \right| \tag{21.27}$$

where we have used an absolute value sign because amplitudes must be positive. Depending upon the phase difference of the two waves, the amplitude of their superposition can be anywhere from zero (perfect destructive interference) to $2a$ (maximum constructive interference).

The amplitude has its maximum value $A = 2a$ if $\cos(\Delta\phi/2) = \pm 1$. This occurs when

$$\Delta\phi = 2m\pi \qquad \text{(maximum amplitude } A = 2a) \tag{21.28}$$

where m is an integer. Similarly, the amplitude is zero if $\cos(\Delta\phi/2) = 0$, which occurs when

$$\Delta\phi = 2\left(m + \frac{1}{2}\right)\pi \qquad \text{(minimum amplitude } A = 0) \tag{21.29}$$

Equations 21.28 and 21.29 are identical to the conditions of Equations 21.22 and 21.23 for constructive and destructive interference. We initially found these conditions by considering the alignment of the crests and troughs. Now we have confirmed them with an algebraic addition of the waves.

It is entirely possible, of course, that the two waves are neither exactly in phase nor exactly out of phase. Equation 21.27 allows us to calculate the amplitude of the superposition for any value of the phase difference. As an example, Figure 21.22 shows the calculated interference of two waves that differ in phase by 40°, by 90°, and by 160°.

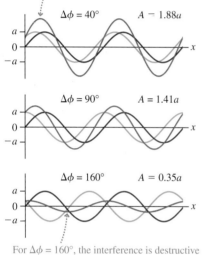

For $\Delta\phi = 40°$, the interference is constructive but not maximum constructive.

$\Delta\phi = 40°$ $A = 1.88a$

$\Delta\phi = 90°$ $A = 1.41a$

$\Delta\phi = 160°$ $A = 0.35a$

For $\Delta\phi = 160°$, the interference is destructive but not perfect destructive.

FIGURE 21.22 The interference of two waves for three different values of the phase difference.

EXAMPLE 21.9 More interference of sound waves

Two loudspeakers emit 500 Hz sound waves with an amplitude of 0.10 mm. Speaker 2 is 1.00 m behind speaker 1, and the phase difference between the speakers is 90°. What is the amplitude of the sound wave at a point 2.00 m in front of speaker 1?

MODEL The amplitude is determined by the interference of the two waves. Assume that the speed of sound has its room-temperature value of 343 m/s.

SOLVE The amplitude of the sound wave is

$$A = \left| 2a\cos(\Delta\phi/2) \right|$$

where $a = 0.10$ mm and the phase difference between the waves is

$$\Delta\phi = \phi_2 - \phi_1 = 2\pi\frac{\Delta x}{\lambda} + \Delta\phi_0$$

The sound's wavelength is

$$\lambda = \frac{v}{f} = \frac{343 \text{ m/s}}{500 \text{ Hz}} = 0.686 \text{ m}$$

Distances $x_1 = 2.0$ m and $x_2 = 3.0$ m are measured from the speakers, so the path-length difference is $\Delta x = 1.00$ m. We're given that the inherent phase difference between the speakers is $\Delta\phi_0 = \pi/2$ rad. Thus the phase difference at the observation point is

$$\Delta\phi = 2\pi\frac{\Delta x}{\lambda} + \Delta\phi_0 = 2\pi\frac{1.00 \text{ m}}{0.686 \text{ m}} + \frac{\pi}{2} \text{ rad} = 10.73 \text{ rad}$$

and the amplitude of the wave at this point is

$$A = \left| 2a\cos\left(\frac{\Delta\phi}{2}\right) \right| = (0.200 \text{ mm})\cos\left(\frac{10.73}{2}\right) = 0.121 \text{ mm}$$

ASSESS The interference is constructive, because $A > a$, but less than maximum constructive interference.

Application: Thin-Film Optical Coatings

Interference in one dimension has an important application in the optics industry, namely the use of **thin-film optical coatings.** These films, less than 1 μm (10^{-6} m) thick, are placed on glass surfaces, such as lenses, to control reflections from the glass. Dichroic glass, shown in the photograph that opens this chapter, is made with a thin-film coating on the glass. Thin films are also used for the anti-reflection coatings used on the lenses in cameras, microscopes, and other optical equipment.

Figure 21.23 shows a light wave of wavelength λ approaching a piece of glass that is coated with a transparent film whose index of refraction is n. The thickness d of this film has been greatly exaggerated in the figure. The air-film boundary is a discontinuity at which the wave speed suddenly decreases, and you saw earlier, in Figure 21.8, that a discontinuity causes a reflection. Most of the light is transmitted into the film, but a little bit is reflected.

Furthermore, you saw in Figure 21.8 that the wave reflected from a discontinuity at which the speed decreases is *inverted* with respect to the incident wave. For a sinusoidal wave, which we're now assuming, the inversion is represented mathematically as phase shift of π rad. The speed of a light wave decreases when it enters a material with a *larger* index of refraction. Thus **a light wave that reflects from a boundary at which the index of refraction increases has a phase shift of π rad.** There is no phase shift for the reflection from a boundary at which the index of refraction decreases. The reflection in Figure 21.23 is from a boundary between air ($n_{air} = 1.00$) and a transparent film with $n_{film} > n_{air}$, so the reflected wave is inverted due to the phase shift of π rad.

When the transmitted wave reaches the glass, most of it continues on into the glass but a portion is reflected back to the left. We'll assume that the index of refraction of the glass is larger than that of the film, $n_{glass} > n_{film}$, so this reflection also has a phase shift of π rad. This second reflection, after traveling back through the film, passes back into the air. There are now *two* equal-frequency waves traveling to the left, and these waves will interfere. If the two reflected waves are *in phase,* they will interfere constructively to cause a *strong reflection.* If the two reflected waves are *out of phase,* they will interfere destructively to cause a *weak reflection* or, if their amplitudes are equal, *no reflection* at all.

This suggests practical uses for thin-film optical coatings. The reflections from glass surfaces, even if weak, are often undesirable. For example, reflections degrade the performance of optical equipment. These reflections can be eliminated by coating the glass with a film whose thickness is chosen to cause *destructive* interference of the two reflected waves. This is called an *antireflection coating.*

Dichroic glass is created by applying a coating that causes strong *constructive* interference for one wavelength or one band of wavelengths. This produces the color you see in reflected light. To conserve energy, any light that is strongly reflected must be absent from the light that is transmitted through the glass. Consequently, the transmitted light has the complementary color—white light minus the reflected color.

The phase difference between the two reflected waves is

$$\Delta\phi = \phi_2 - \phi_1 = (kx_2 + \phi_{20} + \pi \text{ rad}) - (kx_1 + \phi_{10} + \pi \text{ rad})$$
$$= 2\pi\frac{\Delta x}{\lambda_f} - \Delta\phi_0 \qquad (21.30)$$

where we explicitly included the reflection phase shift of each wave. In this case, because *both* waves had a phase shift of π rad, the reflection phase shifts cancel.

The wavelength λ_f is the wavelength *in the film* because that's where the path-length difference Δx occurs. You learned in Chapter 20 that the wavelength in a

The colors of an oil slick are due to thin-film interference. Different wavelengths of light, and thus different colors, are strongly reflected as the thickness of the oil film slowly changes.

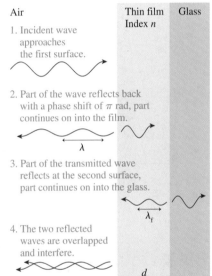

FIGURE 21.23 The two reflections, one from the coating and one from the glass, interfere.

transparent material with index of refraction n is $\lambda_f = \lambda/n$, where the unsubscripted λ is the wavelength in vacuum or air. That is, λ is the wavelength that we measure on "our" side of the air-film boundary.

The path-length difference between the two waves is $\Delta x = 2d$ because wave 2 travels through the film *twice* before rejoining wave 1. The two waves have a common origin—the initial division of the incident wave at the front surface of the film—so the inherent phase difference is $\Delta\phi_0 = 0$. Thus the phase difference of the two reflected waves is

$$\Delta\phi = 2\pi\frac{2d}{\lambda/n} = 2\pi\frac{2nd}{\lambda} \qquad (21.31)$$

The interference is constructive, causing a strong reflection, when $\Delta\phi = 2m\pi$ rad. Constructive interference occurs for wavelengths

$$\lambda_C = \frac{2nd}{m} \qquad m = 1, 2, 3, \ldots \qquad \text{(constructive interference)} \quad (21.32)$$

You will notice that m starts with 1, rather than 0, in order to give meaningful results. Destructive interference, with minimum reflection, requires $\Delta\phi = 2\left(m - \frac{1}{2}\right)\pi$ rad. This occurs for wavelengths

$$\lambda_D = \frac{2nd}{m - \frac{1}{2}} \qquad m = 1, 2, 3, \ldots \qquad \text{(destructive interference)} \quad (21.33)$$

We've used $m - \frac{1}{2}$, rather than $m + \frac{1}{2}$, so that m can start with 1 to match the condition for constructive interference.

NOTE ▶ The exact condition for constructive or destructive interference is satisfied for only a few discrete wavelengths λ. Nonetheless, reflections are strongly enhanced (nearly constructive interference) for a range of wavelengths near λ_C. Likewise, there is a range of wavelengths near λ_D for which the reflection is nearly canceled. ◀

EXAMPLE 21.10 **Designing an antireflection coating**
Magnesium fluoride (MgF_2) is often used as an antireflection coating on lenses. The index of refraction of MgF_2 is 1.39. What is the thinnest film of MgF_2 that works as an antireflection coating at $\lambda = 510$ nm, near the center of the visible spectrum?

MODEL Reflection is minimized if the two reflected waves interfere destructively.

SOLVE The film thicknesses that cause destructive interference at wavelength λ are

$$d = \left(m - \frac{1}{2}\right)\frac{\lambda}{2n}$$

The thinnest film has $m = 1$. Its thickness is

$$d = \frac{\lambda}{4n} = \frac{510 \text{ nm}}{4(1.39)} = 92 \text{ nm}$$

The film thickness is significantly less than the wavelength of visible light!

ASSESS The reflected light is completely eliminated (perfect destructive interference) only if the two reflected waves have equal amplitudes. In practice, they don't. Nonetheless, the reflection is reduced from ≈4% of the incident intensity for "bare glass" to well under 1%. Furthermore, the intensity of reflected light is much reduced across most of the visible spectrum (400–700 nm), even though the phase difference deviates more and more from π rad as the wavelength moves away from 510 nm. It is the increasing reflection at the ends of the visible spectrum ($\lambda \approx 400$ nm and $\lambda \approx 700$ nm), where $\Delta\phi$ deviates significantly from π rad, that gives a reddish-purple tinge to the lenses on cameras and binoculars.

Homework problems will let you "design" a piece of dichroic glass and explore situations where only one of the two reflections has a reflection phase shift of π rad.

21.7 Interference in Two and Three Dimensions

Ripples on a lake move in two dimensions. The glow from a light bulb spreads outward as a spherical wave. In Section 20.4, we noted that a circular or spherical wave can be written

$$D(r, t) = a\sin(kr - \omega t + \phi_0) \qquad (21.34)$$

where r is the distance measured outward from the source. Equation 21.34 is our familiar wave equation with the one-dimensional coordinate x replaced by a more general radial coordinate r. Strictly speaking, the amplitude a of a circular or spherical wave diminishes as r increases. However, we will assume that a remains essentially constant over the region in which we study the wave. Figure 21.24 shows the wave-front diagram for a circular or spherical wave. Recall that the wave fronts represent the *crests* of the wave and are spaced by the wavelength λ.

What happens when two circular or spherical waves overlap? For example, imagine two paddles oscillating up and down on the surface of a pond. We will assume that the two paddles oscillate with the same frequency and amplitude and that they are in phase. Figure 21.25 shows the wave fronts of the two waves. The ripples overlap as they travel, and, as was the case in one dimension, this causes interference.

Constructive interference with $A = 2a$ occurs where two crests align or two troughs align. Several locations of constructive interference are marked in Figure 21.25. Intersecting wave fronts are points where two crests are aligned. It's a bit harder to visualize, but two troughs are aligned when a midpoint between two wave fronts is overlapped with another midpoint between two wave fronts. Destructive interference with $A = 0$ occurs where the crest of one wave aligns with a trough of the other wave. Several points of destructive interference are also indicated in Figure 21.25.

A picture on a page is static, but the wave fronts are in motion. Try to imagine the wave fronts of Figure 21.25 expanding outward as new circular rings are born at the sources. The waves will move forward half a wavelength during half a period, causing the crests in Figure 21.25 to be replaced by troughs while the troughs become crests.

The important point to recognize is that **the motion of the waves does not affect the points of constructive and destructive interference.** Points in the figure where two crests overlap will become points where two troughs overlap, but this overlap is still constructive interference. Similarly, points in the figure where a crest and a trough overlap will become a point where a trough and a crest overlap—still destructive interference.

The mathematical description of interference in two or three dimensions is very similar to that of one-dimensional interference. The net displacement of a particle in the medium is

$$D = D_1 + D_2 = a\sin(kr_1 - \omega t + \phi_{10}) + a\sin(kr_2 - \omega t + \phi_{20}) \quad (21.35)$$

The only difference between Equation 21.35 and the earlier one-dimensional Equation 21.24 is that the linear coordinates x_1 and x_2 have been changed to radial coordinates r_1 and r_2. Thus our conclusions are unchanged. The superposition of the two waves yields a wave traveling outward with amplitude

$$A = \left| 2a\cos\left(\frac{\Delta\phi}{2}\right) \right| \qquad (21.36)$$

where the phase difference, with x replaced by r, is now

$$\Delta\phi = 2\pi\frac{\Delta r}{\lambda} + \Delta\phi_0 \qquad (21.37)$$

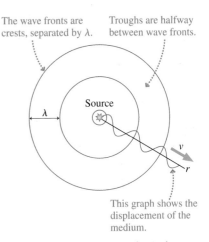

The wave fronts are crests, separated by λ.

Troughs are halfway between wave fronts.

Source

λ

v

r

This graph shows the displacement of the medium.

FIGURE 21.24 A circular or spherical wave.

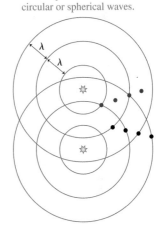

Two in-phase sources emit circular or spherical waves.

λ

λ

● Points of constructive interference. A crest is aligned with a crest, or a trough with a trough.

● Points of destructive interference. A crest is aligned with a trough of another wave.

FIGURE 21.25 The overlapping ripple patterns of two sources. A few points of constructive and destructive interference are noted.

Two overlapping water waves create an interference pattern.

The term $2\pi(\Delta r/\lambda)$ is the phase difference that arises when the waves travel different distances from the sources to the point at which they combine. Δr itself is the *path-length difference*. As before, $\Delta\phi_0$ is any inherent phase difference of the sources themselves.

Maximum constructive interference with $A = 2a$ occurs, just as in one dimension, at those points where $\cos(\Delta\phi/2) = \pm 1$. Similarly, perfect destructive interference occurs at points where $\cos(\Delta\phi/2) = 0$. The conditions for constructive and destructive interference are

Maximum constructive interference:
$$\Delta\phi = 2\pi\frac{\Delta r}{\lambda} + \Delta\phi_0 = 2m\pi$$

Perfect destructive interference:
$$\Delta\phi = 2\pi\frac{\Delta r}{\lambda} + \Delta\phi_0 = 2\left(m + \frac{1}{2}\right)\pi$$

$$m = 0, 1, 2, \ldots \quad (21.38)$$

For two identical sources (i.e., sources that oscillate in phase with $\Delta\phi_0 = 0$), the conditions for constructive and destructive interference are very simple:

Constructive: $\Delta r = m\lambda$

Destructive: $\Delta r = \left(m + \dfrac{1}{2}\right)\lambda$

(identical sources)   (21.39)

The waves from two identical sources interfere constructively at points where the path-length difference is an integer number of wavelengths because, for these values of Δr, crests are aligned with crests and troughs with troughs. **The waves interfere destructively where the path-length difference is a half-integer number of wavelengths** because, for these values of Δr, crests are aligned with troughs. These two statements are the essence of interference.

NOTE ▶ Equation 21.39 applies only if the sources are in phase. If the sources are not in phase, you must use the more general Equation 21.38 to locate the points of constructive and destructive interference. ◀

Wave fronts are spaced exactly one wavelength apart, hence we can measure the distances r_1 and r_2 simply by counting the rings in the wave-front pattern. In Figure 21.26, which is based on Figure 21.25, Point A is distance $r_1 = 3\lambda$ from the first source and $r_2 = 2\lambda$ from the second. The path-length difference is $\Delta r_A = 1\lambda$, the condition for the maximum constructive interference of identical sources. Point B, on the other hand, has $\Delta r_B = \frac{1}{2}\lambda$, so it is a point of perfect destructive interference.

NOTE ▶ Interference is determined by Δr, the path-length *difference,* rather than by r_1 or r_2. ◀

• At A, $\Delta r_A = \lambda$, so this is a point of constructive interference.

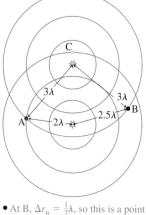

• At B, $\Delta r_B = \frac{1}{2}\lambda$, so this is a point of destructive interference.

FIGURE 21.26 The path-length difference Δr determines whether the interference at a particular point is constructive or destructive.

STOP TO THINK 21.5   The interference at point C in Figure 21.26 is

a. Maximum constructive.
b. Constructive, but less than maximum.
c. Perfect destructive.
d. Destructive, but not perfect.
e. There is no interference at point C.

We can now locate the points of maximum constructive interference, for which $\Delta r = m\lambda$, by drawing a line through *all* the points at which $\Delta r = 0$, another line through all the points at which $\Delta r = \lambda$, and so on. These lines, shown as red lines in Figure 21.27, are called **antinodal lines.** They are analogous to the antinodes of a standing wave, hence the name. An antinode is a *point* of maximum constructive interference; for circular waves, oscillation at maximum amplitude occurs along a continuous *line*. Similarly, destructive interference occurs along lines called **nodal lines.** The displacement is *always zero* along these lines, just as it is at a node in a standing-wave pattern.

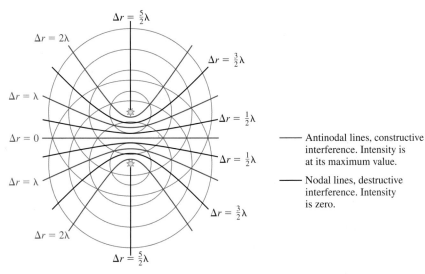

FIGURE 21.27 The points of constructive and destructive interference fall along antinodal and nodal lines.

A Problem-Solving Strategy for Interference Problems

The information in this section is the basis of a strategy for solving interference problems. This strategy applies equally well to interference in one dimension if you use Δx instead of Δr.

(MP) PROBLEM-SOLVING STRATEGY 21.1 **Interference of two waves**

MODEL Make simplifying assumptions, such as assuming waves are circular and of equal amplitude.

VISUALIZE Draw a picture showing the sources of the waves and the point where the waves interfere. Give relevant dimensions. Identify the distances r_1 and r_2 from the sources to the point. Note any phase difference $\Delta\phi_0$ between the two sources.

SOLVE The interference depends on the path-length difference $\Delta r = r_2 - r_1$ and the source phase difference $\Delta\phi_0$.

Constructive: $\Delta\phi = 2\pi\dfrac{\Delta r}{\lambda} + \Delta\phi_0 = 2m\pi$

$$m = 0, 1, 2, \ldots$$

Destructive: $\Delta\phi = 2\pi\dfrac{\Delta r}{\lambda} + \Delta\phi_0 = 2\left(m + \dfrac{1}{2}\right)\pi$

For identical sources ($\Delta\phi_0 = 0$), the interference is constructive if $\Delta r = m\lambda$, destructive if $\Delta r = (m + \frac{1}{2})\lambda$.

ASSESS Check that your result has the correct units, is reasonable, and answers the question.

EXAMPLE 21.11 Two-dimensional interference between two loudspeakers

Two loudspeakers are 2.0 m apart and in phase with each other. Both emit 700 Hz sound waves into a room where the speed of sound is 341 m/s. A listener stands 5.0 m in front of the loudspeakers and 2.0 m to one side of the center. Is the interference at this point constructive, destructive, or something in between? How will the situation differ if the loudspeakers are out of phase?

MODEL The two speakers are sources of in-phase, circular waves. The overlap of these waves causes interference.

VISUALIZE Figure 21.28 shows the loudspeakers and defines the distances r_1 and r_2 to the point of observation. The figure includes dimensions and notes that $\Delta\phi_0 = 0$ rad.

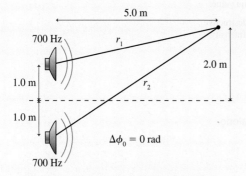

FIGURE 21.28 Pictorial representation of the interference between two loudspeakers.

SOLVE It's not r_1 or r_2 that matter, but the *difference* Δr between them. From the geometry of the figure we can calculate that

$$r_1 = \sqrt{(5.0\text{ m})^2 + (1.0\text{ m})^2} = 5.10\text{ m}$$
$$r_2 = \sqrt{(5.0\text{ m})^2 + (3.0\text{ m})^2} = 5.83\text{ m}$$

Thus the path-length difference is $\Delta r = r_2 - r_1 = 0.73$ m. The wavelength of the sound waves is

$$\lambda = \frac{v}{f} = \frac{341\text{ m/s}}{700\text{ Hz}} = 0.487\text{ m}$$

In terms of wavelengths, the path-length difference is $\Delta r/\lambda = 1.50$, or

$$\Delta r = \frac{3}{2}\lambda$$

Because the sources are in phase ($\Delta\phi_0 = 0$), this is the condition for *destructive* interference. If the sources were out of phase ($\Delta\phi_0 = \pi$ rad), then the phase difference of the waves at the listener would be

$$\Delta\phi = 2\pi\frac{\Delta r}{\lambda} + \Delta\phi_0 = 2\pi\left(\frac{3}{2}\right) + \pi\text{ rad} = 4\pi\text{ rad}$$

This is an integer multiple of 2π rad, so in this case the interference would be *constructive*.

ASSESS Both the path-length difference *and* any inherent phase difference of the sources must be considered when evaluating interference.

Picturing Interference

A *contour map* is a useful way to visualize an interference pattern. Figure 21.29a shows the superposition of the waves from two identical sources ($\Delta\phi_0 = 0$) emitting waves with $\lambda = 1$ m. The sources, indicated with black dots, are located at $y = \pm 1$ m. Positive displacements are shown in red, with the deepest red representing the maximum displacement of the wave at this instant in time. These are the points where the crests of the individual waves interfere constructively to give $D = 2a$. Negative displacements are blue, with the darkest blue being the most negative displacement of the wave. These are also points of constructive interference, with two troughs overlapping to give $D = -2a$.

To understand this figure, try to visualize the waves expanding outward from center. The red-blue-red-blue-red-. . . pattern of crests and troughs moves outward along the antinodal lines as a *traveling wave* of amplitude $A = 2a$. Nothing ever happens along the nodal lines, where the amplitude is always zero. Between the nodal and antinodal lines are traveling waves whose amplitude varies between 0 and $2a$.

Suppose you were to observe the *intensity* of the wave as it crosses the vertical line at $x = 4$ m on the right edge of the figure. If, for example, these are sound waves, you could listen to (or measure, with a microphone) the sound intensity as you walk from $(x, y) = (4\text{ m}, -4\text{ m})$ at the bottom of the figure to $(x, y) = (4\text{ m}, 4\text{ m})$ at the top. The intensity is zero as you cross the nodal lines at $y \approx \pm 1$ m ($\Delta r = \frac{1}{2}\lambda$). The intensity is maximum at the antinodal lines at

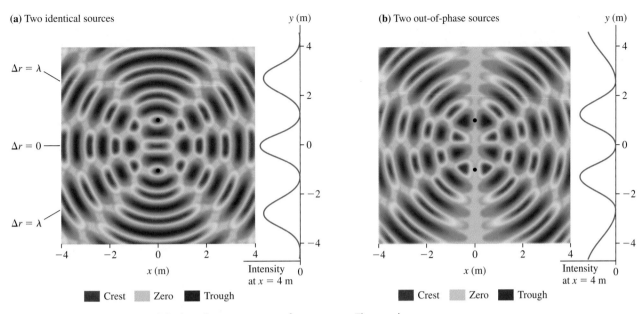

(a) Two identical sources

$\Delta r = \lambda$

$\Delta r = 0$

$\Delta r = \lambda$

x (m)

y (m)

Intensity at *x* = 4 m

■ Crest ■ Zero ■ Trough

(b) Two out-of-phase sources

x (m)

y (m)

Intensity at *x* = 4 m

■ Crest ■ Zero ■ Trough

FIGURE 21.29 A contour map of the interference pattern of two sources. The graph on the right side of each figure shows the wave intensity along a vertical line at *x* = 4 m.

$y = 0$ ($\Delta r = 0$) and $y \approx \pm 2.5$ m ($\Delta r = \lambda$), where a wave of maximum amplitude streams out from the sources.

The intensity is shown in the rather unusual graph on the right side of Figure 21.29a. It is unusual in the sense that the intensity, the quantity of interest, is graphed to the left. The peaks are the points of constructive interference, where you would measure maximum amplitude. The zeros are points of destructive interference, where the intensity is zero.

Figure 21.29b is a contour map of the interference pattern produced by the same two sources but with the sources themselves now out of phase ($\Delta \phi_0 = \pi$ rad). We'll leave the investigation of this figure for you to study, but notice that the nodal and antinodal lines are reversed from those of Figure 21.29a.

EXAMPLE 21.12 The intensity of two interfering loudspeakers

Two loudspeakers are 6.0 m apart and in phase. They emit equal-amplitude sound waves with a wavelength of 1.0 m. Each speaker alone creates sound with intensity I_0. An observer at point A is 10 m in front of the plane containing the two loudspeakers and centered between them. A second observer at point B is 10 m directly in front of one of the speakers. In terms of I_0, what is the intensity I_A at point A and the intensity I_B at point B?

MODEL The two speakers are sources of in-phase, circular waves. The overlap of these waves causes interference.

VISUALIZE Figure 21.30 shows the two loudspeakers and the two points of observation. Distances r_1 and r_2 are defined for point B.

SOLVE Let the amplitude of the wave from each speaker be a. The intensity of a wave is proportional to the square of the amplitude, so the intensity of each speaker alone is $I_0 = ca^2$,

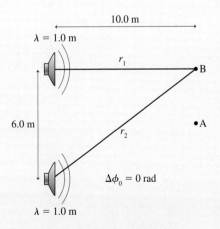

10.0 m

$\lambda = 1.0$ m

r_1

B

6.0 m

r_2

•A

$\Delta \phi_0 = 0$ rad

$\lambda = 1.0$ m

FIGURE 21.30 Pictorial representation of the interference between two loudspeakers.

where c is an unknown proportionality constant. Point A is a point of constructive interference because the speakers are in phase ($\Delta\phi_0 = 0$) and the path-length difference is $\Delta r = 0$. The amplitude at this point is given by Equation 21.36:

$$A_A = \left|2a\cos\left(\frac{\Delta\phi}{2}\right)\right| = 2a\cos(0) = 2a$$

Consequently, the intensity at this point is

$$I_A = cA_A^2 = c(2a)^2 = 4ca^2 = 4I_0$$

The intensity at A is four times that of either speaker played alone. At point B, the path-length difference is

$$\Delta r = \sqrt{(10.0\text{ m})^2 + (6.0\text{ m})^2} - 10.0\text{ m} = 1.662\text{ m}$$

The phase difference of the waves at this point is

$$\Delta\phi = 2\pi\frac{\Delta r}{\lambda} = 2\pi\frac{1.662\text{ m}}{1.0\text{ m}} = 10.44\text{ rad}$$

Consequently, the amplitude at B is

$$A_B = \left|2a\cos\left(\frac{\Delta\phi}{2}\right)\right| = |2a\cos(5.22\text{ rad})| = 0.972a$$

Thus the intensity at this point is

$$I_B = cA_B^2 = c(0.972a)^2 = 0.95ca^2 = 0.95I_0$$

ASSESS Although B is directly in front of one of the speakers, superposition of the two waves results in an intensity that is less than it would be if this speaker played alone.

STOP TO THINK 21.6   These two loudspeakers are in phase. They emit equal-amplitude sound waves with a wavelength of 1.0 m. At the point indicated, is the interference maximum constructive, perfect destructive, or something in between?

8.5 m

9.5 m

$\lambda = 1.0$ m

$\lambda = 1.0$ m

21.8 Beats

10.7

Thus far we have looked at the superposition of sources having the same wavelength and frequency. We can also use the principle of superposition to investigate a phenomenon that is easily demonstrated with two sources of slightly different frequency.

If you listen to two sounds with very different frequencies, such as a high note and a low note, you hear two distinct tones. But if the frequency difference is very small, just one or two hertz, then you hear a single tone whose intensity is *modulated* once or twice every second. That is, the sound goes up and down in volume, loud, soft, loud, soft, . . . , making a distinctive sound pattern called **beats.**

Consider two sinusoidal waves traveling along the x-axis with angular frequencies $\omega_1 = 2\pi f_1$ and $\omega_2 = 2\pi f_2$. The two waves are

$$D_1 = a\sin(k_1 x - \omega_1 t + \phi_{10})$$
$$D_2 = a\sin(k_2 x - \omega_2 t + \phi_{20})$$

(21.40)

where the subscripts 1 and 2 indicate that the frequencies, wave numbers, and phase constants of the two waves may be different.

To simplify the analysis, let's make several assumptions:

1. The two waves have the same amplitude a,
2. A detector, such as your ear, is located at the origin ($x = 0$),
3. The two sources are in phase ($\phi_{10} = \phi_{20}$), and
4. The source phases happen to be $\phi_{10} = \phi_{20} = \pi$ rad.

None of these assumptions is essential to the outcome. All could be otherwise and we would still come to basically the same conclusion, but the mathematics would be far messier. Making these assumptions allows us to emphasize the physics with the least amount of mathematics.

With these assumptions, the two waves as they reach the detector at $x = 0$ are

$$D_1 = a\sin(-\omega_1 t + \pi) = a\sin\omega_1 t$$
$$D_2 = a\sin(-\omega_2 t + \pi) = a\sin\omega_2 t$$
(21.41)

where we've used the trigonometric identity $\sin(\pi - \theta) = \sin\theta$. The principle of superposition tells us that the *net* displacement of the medium at the detector is the sum of the displacements of the individual waves. Thus

$$D = D_1 + D_2 = a(\sin\omega_1 t + \sin\omega_2 t)$$
(21.42)

Earlier, in our analysis of the mathematics of interference, we used the trigonometric identity

$$\sin\alpha + \sin\beta = 2\cos\left[\frac{1}{2}(\alpha - \beta)\right]\sin\left[\frac{1}{2}(\alpha + \beta)\right]$$

We can use this identity again to write Equation 21.42 as

$$D = 2a\cos\left[\frac{1}{2}(\omega_1 - \omega_2)t\right]\sin\left[\frac{1}{2}(\omega_1 + \omega_2)t\right]$$
$$= \left[2a\cos(\omega_{\mathrm{mod}}t)\right]\sin(\omega_{\mathrm{avg}}t)$$
(21.43)

where $\omega_{\mathrm{avg}} = \frac{1}{2}(\omega_1 + \omega_2)$ is the *average* angular frequency and $\omega_{\mathrm{mod}} = \frac{1}{2}(\omega_1 - \omega_2)$ is called the *modulation frequency*.

We are interested in the situation when the two frequencies are very nearly equal: $\omega_1 \approx \omega_2$. In that case, ω_{avg} hardly differs from either ω_1 or ω_2 while ω_{mod} is very near to—but not exactly—zero. When ω_{mod} is very small, the term $\cos(\omega_{\mathrm{mod}}t)$ oscillates *very* slowly. We have grouped it with the $2a$ term because, together, they provide a slowly changing "amplitude" for the rapid oscillation at frequency ω_{avg}.

Figure 21.31 is a history graph of the wave at the detector ($x = 0$). It shows the oscillation of the air against your ear drum at frequency $f_{\mathrm{avg}} = \omega_{\mathrm{avg}}/2\pi = \frac{1}{2}(f_1 + f_2)$. This oscillation determines the note you hear; it differs little from the two notes at frequencies f_1 and f_2. We are especially interested in the time-dependent amplitude, shown as a dotted line, that is given by the term $2a\cos(\omega_{\mathrm{mod}}t)$. This periodically varying amplitude is called a **modulation** of the wave, which is where ω_{mod} gets its name.

As the amplitude rises and falls, the sound alternates as loud, soft, loud, soft, and so on. But that is exactly what you hear when you listen to beats! The alternating loud and soft sounds arise from the two waves being alternately in phase and out of phase, causing constructive and then destructive interference.

Imagine two people walking side-by-side at just slightly different paces. Initially both of their right feet hit the ground together, but after a while they get out of step. A little bit later they are back in step and the process alternates. The sound waves are doing the same. Initially the crests of each wave, of amplitude a, arrive together at your ear and the net displacement is doubled to $2a$. But after a while the two waves, being of slightly different frequency, get out of step and a crest of

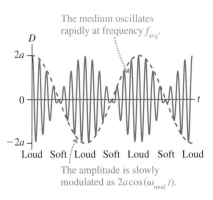

The medium oscillates rapidly at frequency f_{avg}.

The amplitude is slowly modulated as $2a\cos(\omega_{\mathrm{mod}}t)$.

FIGURE 21.31 Beats are caused by the superposition of two waves of nearly identical frequency.

one arrives with a trough of the other. When this happens, the two waves cancel each other to give a net displacement of zero. This process alternates over and over, loud and soft.

Notice, from the figure, that the sound intensity rises and falls *twice* during one cycle of the modulation envelope. Each "loud-soft-loud" is one beat, so the **beat frequency** f_{beat}, which is the number of beats per second, is *twice* the modulation frequency $f_{mod} = \omega_{mod}/2\pi$. From the above definition of ω_{mod}, the beat frequency is

$$f_{beat} = 2f_{mod} = 2\frac{\omega_{mod}}{2\pi} = 2 \cdot \frac{1}{2}\left(\frac{\omega_1}{2\pi} - \frac{\omega_2}{2\pi}\right) = f_1 - f_2 \qquad (21.44)$$

where, to keep f_{beat} from being negative, we will always let f_1 be the larger of the two frequencies. The beat frequency is simply the *difference* between the two individual frequencies.

EXAMPLE 21.13 Listening to beats
One flutist plays a note of 510 Hz while a second plays a note of 512 Hz. What frequency do you hear? What is the beat frequency?

SOLVE You hear a note with frequency

$$f_{avg} = 511 \text{ Hz}$$

The beat frequency is

$$f_{beat} = f_1 - f_2 = 2 \text{ Hz}$$

You (and they) would hear two beats per second.

ASSESS If a 510 Hz note and a 512 Hz note were played separately, you would not be able to perceive the slight difference in frequency. But when the two notes are played together, the quite obvious beats tell you that the frequencies are slightly different. Musicians learn to make constant minor adjustments in their tuning as they play in order to eliminate beats between themselves and other players.

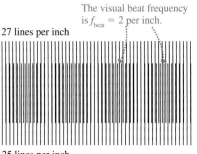

The visual beat frequency is $f_{beat} = 2$ per inch.

27 lines per inch

25 lines per inch

FIGURE 21.32 A graphical example of beats.

Beats aren't limited to sound waves. Figure 21.32 shows a graphical example of beats. Two "fences" of slightly different frequencies—25 lines per inch and 27 lines per inch—are superimposed on each other. The density of the lines varies as they alternate in step and out of step, giving a visual "loud/soft" alternation. The difference in the two frequencies is two lines per inch. You can confirm, with a ruler, that the figure has two "beats" per inch, in agreement with Equation 21.44.

Beats are important in many other situations. For example, you have probably seen movies where rotating wheels seem to turn slowly backward. Why is this? Suppose the movie camera is shooting at 30 frames per second but the wheel is rotating 32 times per second. The combination of the two produces a "beat" of 2 Hz, meaning that the wheel appears to rotate only twice per second. The same is true if the wheel is rotating 28 times per second, but in this case, where the wheel frequency slightly lags the camera frequency, it appears to rotate *backward* twice per second!

STOP TO THINK 21.7 You hear three beats per second when two sound tones are generated. The frequency of one tone is known to be 610 Hz. The frequency of the other is

a. 604 Hz.   b. 607 Hz.
c. 613 Hz.   d. 616 Hz.
e. Either a or d.   f. Either b or c.

SUMMARY

The goal of Chapter 21 has been to understand and use the idea of superposition.

GENERAL PRINCIPLES

Principle of Superposition

The displacement of a medium when more than one wave is present is the sum of the displacements due to each individual wave.

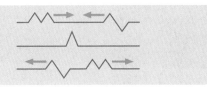

IMPORTANT CONCEPTS

Standing waves are due to the superposition of two traveling waves moving in opposite directions.

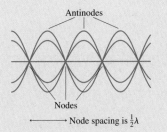

The amplitude at position x is

$$A(x) = 2a \sin kx$$

where a is the amplitude of each wave.

The boundary conditions determine which standing wave frequencies and wavelengths are allowed.

Interference

In general, the superposition of two or more waves into a single wave is called interference.

Maximum constructive interference occurs where crests are aligned with crests and troughs with troughs. These waves are in phase. The maximum displacement is $A = 2a$.

Perfect destructive interference occurs where crests are aligned with troughs. These waves are out of phase. The amplitude is $A = 0$.

Interference depends on the phase difference $\Delta\phi$ between the two waves.

Constructive: $\Delta\phi = 2\pi\dfrac{\Delta r}{\lambda} + \Delta\phi_0 = 2m\pi$

Destructive: $\Delta\phi = 2\pi\dfrac{\Delta r}{\lambda} + \Delta\phi_0 = 2(m + \frac{1}{2})\pi$

Δr is the path-length difference of the two waves and $\Delta\phi_0$ is any phase difference between the sources. For identical sources (in phase, $\Delta\phi_0 = 0$):

Interference is constructive if the path-length difference $\Delta r = m\lambda$.

Interference is destructive if the path-length difference $\Delta r = (m + \frac{1}{2})\lambda$.

The amplitude at a point where the phase difference is $\Delta\phi$ is $A = \left| 2a \cos\left(\dfrac{\Delta\phi}{2}\right) \right|$

Antinodal lines, constructive interference. $A = 2a$

Nodal lines, destructive interference. $A = 0$

APPLICATIONS

Boundary conditions

Strings, electromagnetic waves, and sound waves in closed-closed tubes must have nodes at both ends.

$$\lambda_m = \frac{2L}{m} \qquad f_m = m\frac{v}{2L} = mf_1$$

where $m = 1, 2, 3, \dots$

The frequencies and wavelengths are the same for a sound wave in an open-open tube, which has antinodes at both ends.

A sound wave in an open-closed tube must have a node at the closed end but an antinode at the open end. This leads to

$$\lambda_m = \frac{4L}{m} \qquad f_m = m\frac{v}{4L} = mf_1$$

where $m = 1, 3, 5, 7, \dots$

Beats (loud-soft-loud-soft modulations of intensity) occur when two waves of slightly different frequency are superimposed.

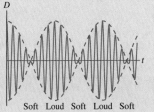

Soft Loud Soft Loud Soft

The beat frequency between waves of frequencies f_1 and f_2 is

$$f_{\text{beat}} = f_1 - f_2$$

TERMS AND NOTATION

principle of superposition	normal mode	thin-film optical coating
standing wave	interference	antinodal line
node	in phase	nodal line
antinode	constructive interference	beats
amplitude function, $A(x)$	out of phase	modulation
boundary condition	destructive interference	beat frequency, f_{beat}
fundamental frequency, f_1	phase difference, $\Delta\phi$	
harmonic	path-length difference, Δx or Δr	

EXERCISES AND PROBLEMS

Exercises

Section 21.1 The Principle of Superposition

1. Figure Ex21.1 is a snapshot graph at $t = 0$ s of two waves approaching each other at 1 m/s. Draw six snapshot graphs, stacked vertically, showing the string at 1 s intervals from $t = 1$ s to $t = 6$ s.

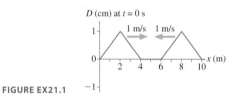

FIGURE EX21.1

2. Figure Ex21.2 is a snapshot graph at $t = 0$ s of two waves approaching each other at 1 m/s. Draw six snapshot graphs, stacked vertically, showing the string at 1 s intervals from $t = 1$ s to $t = 6$ s.

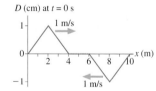

FIGURE EX21.2

3. Figure Ex21.3 is a snapshot graph at $t = 0$ s of two waves approaching each other at 1 m/s. Draw four snapshot graphs, stacked vertically, showing the string at $t = 2, 4, 6$, and 8 s.

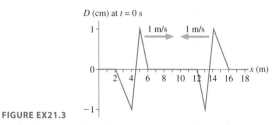

FIGURE EX21.3

4. Figure Ex21.4 is a snapshot graph at $t = 0$ s of two waves approaching each other at 1 m/s. Draw four snapshot graphs, stacked vertically, showing the string at $t = 2, 4, 6$, and 8 s.

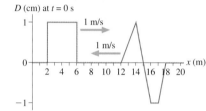

FIGURE EX21.4

5. Figure Ex21.5a is a snapshot graph at $t = 0$ s of two waves approaching each other at 1 m/s.
   a. At what time was the snapshot graph in Figure Ex21.5b taken?
   b. Draw a history graph of the string at $x = 5$ m from $t = 0$ s to $t = 6$ s.

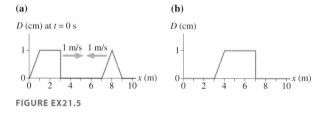

FIGURE EX21.5

Section 21.2 Standing Waves

Section 21.3 Transverse Standing Waves

6. Figure Ex21.6 is a snapshot graph at $t = 0$ s of two waves moving to the right at 1 m/s. The string is fixed at $x = 8$ m. Draw four snapshot graphs, stacked vertically, showing the string at $t = 2, 4, 6$, and 8 s.

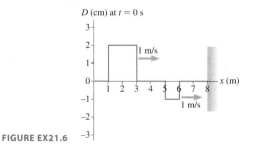

FIGURE EX21.6

7. A 2.0-m-long string is fixed at both ends and tightened until the wave speed is 40 m/s. What is the frequency of the standing wave shown in Figure Ex21.7?

FIGURE EX21.7

60 cm

FIGURE EX21.8

8. Figure Ex21.8 shows a standing wave oscillating at 100 Hz on a string. What is the wave speed?

9. Figure Ex21.9 shows a standing wave that is oscillating at frequency f_0.
   a. How many antinodes will there be if the frequency is doubled to $2f_0$? Explain.
   b. If the tension in the string is increased by a factor of four, for what frequency, in terms of f_0, will the string continue to oscillate as a standing wave with three antinodes?

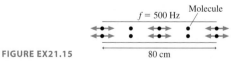

FIGURE EX21.9

10. a. What are the three longest wavelengths for standing waves on a 240-cm-long string that is fixed at both ends?
    b. If the frequency of the second-longest wavelength is 50 Hz, what is the frequency of the third-longest wavelength?

11. Standing waves on a 1.0-m-long string that is fixed at both ends are seen at successive frequencies of 24 Hz and 36 Hz.
    a. What are the fundamental frequency and the wave speed?
    b. Draw the standing-wave pattern when the string oscillates at 36 Hz.

12. A 121-cm-long, 4.0 g string oscillates in its $m = 3$ mode with a frequency of 180 Hz and a maximum amplitude of 5.0 mm. What are (a) the wavelength and (b) the tension in the string?

13. A guitar string with a linear density of 2.0 g/m is stretched between supports that are 60 cm apart. The string is observed to form a standing wave with three antinodes when driven at a frequency of 420 Hz. What are (a) the frequency of the fifth harmonic of this string and (b) the tension in the string?

14. A carbon dioxide laser is an infrared laser. A CO_2 laser with a cavity length of 53.00 cm oscillates in the $m = 100,000$ mode. What are the wavelength and frequency of the laser beam?

Section 21.4 Standing Sound Waves and Musical Acoustics

15. Figure Ex21.15 shows a standing sound wave in an 80-cm-long tube. The tube is filled with an unknown gas. What is the speed of sound in this gas?

$f = 500$ Hz   Molecule

FIGURE EX21.15

80 cm

16. What are the three longest wavelengths for standing sound waves in a 121-cm-long tube that is (a) open at both ends and (b) open at one end, closed at the other?

17. The lowest pedal note on a large pipe organ has a fundamental frequency of 16.4 Hz. This extreme bass note, four octaves below middle C, is more felt than heard. What is the length of pipe between the sounding hole and the open end?

18. The fundamental frequency of an open-open tube is 1500 Hz when the tube is filled with 0°C helium. What is its frequency when filled with 0°C air?

19. A violin string is 30 cm long. It sounds the musical note A (440 Hz) when played without fingering. How far from the end of the string should you place your finger to play the note C (523 Hz)?

20. The lowest note on a grand piano has a frequency of 27.5 Hz. The entire string is 2.00 m long and has a mass of 400 g. The vibrating section of the string is 1.90 m long. What tension is needed to tune this string properly?

Section 21.5 Interference in One Dimension

Section 21.6 The Mathematics of Interference

21. Two loudspeakers in a 20°C room emit 686 Hz sound waves along the x-axis.
    a. If the speakers are in phase, what is the smallest distance between the speakers for which the interference of the sound waves is destructive?
    b. If the speakers are out of phase, what is the smallest distance between the speakers for which the interference of the sound waves is constructive?

22. Two loudspeakers emit sound waves along the x-axis. The sound has maximum intensity when the speakers are 20 cm apart. The sound intensity decreases as the distance between the speakers is increased, reaching zero at a separation of 60 cm.
    a. What is the wavelength of the sound?
    b. If the distance between the speakers continues to increase, at what separation will the sound intensity again be a maximum?

23. Two identical loudspeakers separated by distance d emit 170 Hz sound waves along the x-axis. As you walk along the axis, away from the speakers, you don't hear anything even though both speakers are on. What are three possible values for d? Assume a sound speed of 340 m/s.

24. What is the thinnest film of MgF_2 ($n = 1.39$) on glass that produces a strong reflection for orange light with a wavelength of 600 nm?

25. A very thin oil film ($n = 1.25$) floats on water ($n = 1.33$). What is the thinnest film that produces a strong reflection for green light with a wavelength of 500 nm?

Section 21.7 Interference in Two and Three Dimensions

26. Figure Ex21.26 shows the circular wave fronts emitted by two sources.
    a. Are these sources in phase or out of phase? Explain.
    b. Make a table with rows labeled P, Q, and R and columns labeled r_1, r_2, Δr, and C/D. Fill in the table for points P, Q, and R, giving the distances as multiples of λ and indicating, with a C or a D, whether the interference at that point is constructive or destructive.

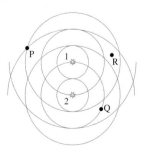

FIGURE EX21.26

27. Figure Ex21.27 shows the circular wave fronts emitted by two sources.
    a. Are these sources in phase or out of phase? Explain.
    b. Make a table with rows labeled P, Q, and R and columns labeled r_1, r_2, Δr, and C/D. Fill in the table for points P, Q, and R, giving the distances as multiples of λ and indicating, with a C or a D, whether the interference at that point is constructive or destructive.

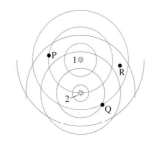

FIGURE EX21.27

28. Two identical loudspeakers 2.0 m apart are emitting 1800 Hz sound waves into a room where the speed of sound is 340 m/s. Is the point 4.0 m in front of one of the speakers, perpendicular to the plane of the speakers, a point of maximum constructive interference, perfect destructive interference, or something in between?

29. Two out-of-phase radio antennas at $x = \pm300$ m on the x-axis are emitting 3.0 MHz radio waves. Is the point $(x, y) = (300\ \text{m}, 800\ \text{m})$ a point of maximum constructive interference, perfect destructive interference, or something in between?

Section 21.8 Beats

30. Two strings are adjusted to vibrate at exactly 200 Hz. Then the tension in one string is increased slightly. Afterward, three beats per second are heard when the strings vibrate at the same time. What is the new frequency of the string that was tightened?

31. A flute player hears four beats per second when she compares her note to a 523 Hz tuning fork (the note C). She can match the frequency of the tuning fork by pulling out the "tuning joint" to lengthen her flute slightly. What was her initial frequency?

Problems

32. Two wave pulses on a string travel in opposite directions at 100 m/s. Figure P21.32 shows a snapshot graph of the string at $t = 0$ s, when the two waves are overlapped, and a snapshot graph of the right-traveling wave at $t = 0.05$ s. Draw a snapshot graph of the left-traveling wave at $t = 0.05$ s.

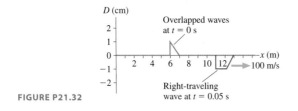

FIGURE P21.32

33. Two waves on a string travel in opposite directions at 100 m/s. Figure P21.33 shows a snapshot graph of the string at $t = 0$ s, when the two waves are overlapped, and a snapshot graph of the left-traveling wave at $t = 0.05$ s. Draw a snapshot graph of the right-traveling wave at $t = 0.05$ s.

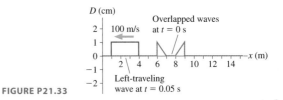

FIGURE P21.33

34. The superposition of two counter-propagating, equal-frequency sinusoidal waves is a standing wave. What about the superposition of two equal-frequency standing waves? Consider the superposition $D = a\sin(kx)\cos(\omega t) + a\cos(kx)\sin(\omega t)$. Each of these two standing waves has period $T = 2\pi/\omega$ and wavelength $\lambda = 2\pi/k$.
    a. Draw nine snapshot graphs, one every eighth of a period from $t = 0$ to $t = T$, showing the x-axis from $x = 0$ to $x = 2\lambda$. Sketch each of the standing waves as a dotted line, then show their superposition as a solid line. Stack the graphs vertically, similar to Figure 21.4a.
    b. Is the superposition a standing wave or a traveling wave? If it's a traveling wave, which way is it moving? Use your graphs to explain.
    c. Use a well-known trigonometric identity to write the displacement D in a form that is clearly a standing wave or a traveling wave. Show that your result agrees with your observations of part b.

35. A 2.0-m-long string vibrates at its second-harmonic frequency with a maximum amplitude of 2.00 cm. One end of the string is at $x = 0$ cm. Find the oscillation amplitude at $x = 10, 20, 30, 40,$ and 50 cm.

36. A string vibrates at its third-harmonic frequency. The amplitude at a point 30 cm from one end is half the maximum amplitude. How long is the string?

37. A string of length L vibrates at its fundamental frequency. The amplitude at a point $\frac{1}{4}L$ from one end is 2.0 cm. What is the amplitude of each of the traveling waves that form this standing wave?

38. Two sinusoidal waves with equal wavelengths travel along a string in opposite directions at 3.0 m/s. The time between two successive instants when the antinodes are at maximum height is 0.25 s. What is the wavelength?

39. An 80-cm-long guitar string with a linear density of 1.0 g/m is under 200 N tension. It is plucked and vibrates at its fundamental frequency. What is the wavelength of the sound wave that reaches your ear in a 20°C room?

40. A violinist places her finger so that the vibrating section of a 1.0 g/m string has a length of 30 cm, then she draws her bow across it. A listener nearby in a 20°C room hears a note with a wavelength of 40 cm. What is the tension in the string?

41. A particularly beautiful note reaching your ear from a rare Stradivarius violin has a wavelength of 39.1 cm. The room is slightly warm, so the speed of sound is 344 m/s. If the string's linear density is 0.60 g/m and the tension is 150 N, how long is the vibrating section of the violin string?

42. A heavy piece of hanging sculpture is suspended by a 90-cm-long, 5.0 g steel wire. When the wind blows hard, the wire hums at its fundamental frequency of 80 Hz. What is the mass of the sculpture?

43. Astronauts visiting Planet X have a 2.5-m-long string whose mass is 5.0 g. They tie the string to a support, stretch it horizontally over a pulley 2.0 m away, and hang a 1.0 kg mass on the free end. Then the astronauts begin to excite standing waves on the string. Their data show that standing waves exist at frequencies of 64 Hz and 80 Hz, but at no frequencies in between. What is the value of g, the acceleration due to gravity, on Planet X?

44. A 75 g bungee cord has an equilibrium length of 1.20 m. The cord is stretched to a length of 1.80 m, then vibrated at 20 Hz. This produces a standing wave with two antinodes. What is the spring constant of the bungee cord?

45. A steel wire is used to stretch a spring. An oscillating magnetic field drives the steel wire back and forth. A standing wave with three antinodes is created when the spring is stretched 8.0 cm. What stretch of the spring produces a standing wave with two antinodes?

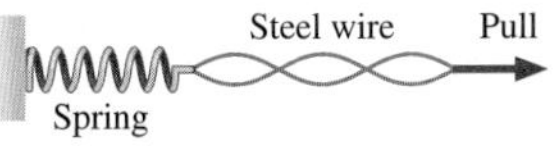

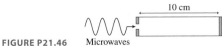

FIGURE P21.45

46. The microwave generator in Figure P21.46 can produce microwaves at any frequency between 10 GHz and 20 GHz. The microwaves are aimed, through a small hole, into a "microwave cavity" that consists of a 10-cm-long cylinder with reflective ends.
    a. Which frequencies will create standing waves in the microwave cavity?
    b. For which of these frequencies is the cavity midpoint an antinode?

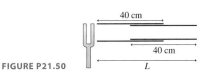

FIGURE P21.46   Microwaves

47. An open-open organ pipe is 78.0 cm long. An open-closed pipe has a fundamental frequency equal to the third harmonic of the open-open pipe. How long is the open-closed pipe?

48. A narrow column of air is found to have standing waves at frequencies of 390 Hz, 520 Hz, and 650 Hz and at no frequencies in between these. The behavior of the tube at frequencies less than 390 Hz or greater than 650 Hz is not known.
    a. Is this an open-open tube or an open-closed tube? Explain.
    b. How long is the tube?
    c. Draw a displacement graph of the 520 Hz standing wave in the tube.
    d. The air in the tube is replaced with carbon dioxide, which has a sound speed of 280 m/s. What are the new frequencies of these three modes?

49. In 1866, the German scientist Adolph Kundt developed a technique for accurately measuring the speed of sound in various gases. A long glass tube, known today as a Kundt's tube, has a vibrating piston at one end and is closed at the other. Very finely ground particles of cork are sprinkled in the bottom of the tube before the piston is inserted. As the vibrating piston is slowly moved forward, there are a few positions that cause the cork particles to collect in small, regularly spaced piles along the bottom. Figure P21.49 shows an experiment in which the tube is filled with pure oxygen and the piston is driven at 400 Hz. What is the speed of sound in oxygen?

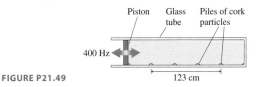

FIGURE P21.49

50. A 40-cm-long tube has a 40-cm-long insert that can be pulled in and out. A vibrating tuning fork is held next to the tube. As the insert is slowly pulled out, the sound from the tuning fork creates standing waves in the tube when the total length L is 42.5 cm, 56.7 cm, and 70.9 cm. What is the frequency of the tuning fork?

FIGURE P21.50

51. A 1.0-m-tall vertical tube is filled with 20°C water. A tuning fork vibrating at 580 Hz is held just over the top of the tube as the water is slowly drained from the bottom. At what water heights, measured from the bottom of the tube, will there be a standing wave in the tube?

52. A 50-cm-long wire with a mass of 1.0 g and a tension of 440 N passes across the open end of an open-closed tube of air. The wire, which is fixed at both ends, is bowed at the center so as to vibrate at its fundamental frequency and generate a sound wave. Then the tube length is adjusted until the fundamental frequency of the tube is heard. What is the length of the tube? Assume $v_{sound} = 340$ m/s.

53. A 25-cm-long wire with a linear density of 20 g/m passes across the open end of an 85-cm-long open-closed tube of air. If the wire, which is fixed at both ends, vibrates at its fundamental frequency, the sound wave it generates excites the second vibrational mode of the tube of air. What is the tension in the wire? Assume $v_{sound} = 340$ m/s.

54. A 50-cm-long wire with a mass of 1.0 g and a tension of 440 N passes across the open top of a vertical tube partially filled with water. The wire, which is fixed at both ends, is bowed at the center so as to vibrate at its fundamental frequency and generate a sound wave. The water level in the tube is slowly lowered until the sound wave from the wire sets up a standing wave in the tube. It is then lowered another 36.0 cm until the next standing wave is detected. Use this information to determine the speed of sound in air.

55. A longitudinal standing wave can be created in a long, thin aluminum rod by stroking the rod with very dry fingers. This is often done as a physics demonstration, creating a high-pitched, very annoying whine. From a wave perspective, the standing wave is equivalent to a sound standing wave in an open-open tube. In particular, both ends of the rod are antinodes. What is the fundamental frequency of a 2.0-m-long aluminum rod?

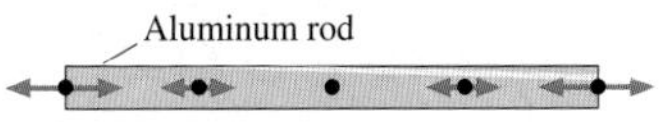

FIGURE P21.55

56. Figure P21.56 is a snapshot picture showing some of the air molecules in a tube of air at their equilibrium positions and when a standing sound wave is present inside the tube.
    a. Is the left end of the tube open or closed? The right end? Explain.
    b. If the tube is 1.5 m long, what is the wavelength of this wave?
    c. Draw a snapshot picture showing the air molecules one half cycle later.

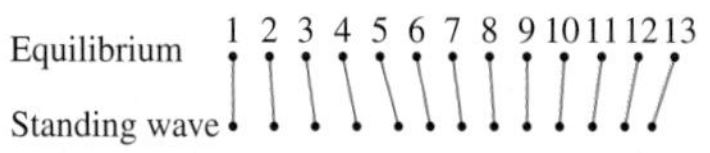

Equilibrium 1 2 3 4 5 6 7 8 9 10 11 12 13

FIGURE P21.56   Standing wave

57. Analyze the standing sound waves in an open-closed tube to show that the possible wavelengths and frequencies are given by Equation 21.18.
58. Two loudspeakers emit sound waves of the same frequency along the x-axis. The amplitude of each wave is a. The sound intensity is minimum when speaker 2 is 10 cm behind speaker 1. The intensity increases as speaker 2 is moved forward and first reaches maximum, with amplitude $2a$, when it is 30 cm in front of speaker 1. What is
    a. The wavelength of the sound?
    b. The phase difference between the two loudspeakers?
    c. The amplitude of the sound if the speakers are placed side by side?
59. Two in-phase loudspeakers emit identical 1000 Hz sound waves along the x-axis. What distance should one speaker be placed behind the other for the sound to have an amplitude 1.5 times that of each speaker alone?
60. Two loudspeakers emit sound waves along the x-axis. Speaker 2 is 2.0 m behind speaker 1. Both loudspeakers are connected to the same signal generator, which is oscillating at 340 Hz, but the wire to speaker 1 passes through a box that delays the signal by 1.47 ms. Is the interference along the x-axis maximum constructive interference, perfect destructive interference, or something in between? Assume $v_{sound} = 340$ m/s.
61. Two loudspeakers emit sound waves along the x-axis. A listener in front of both speakers hears a maximum sound intensity when speaker 2 is at the origin and speaker 1 is at $x = 0.50$ m. If speaker 1 is slowly moved forward, the sound intensity decreases and then increases, reaching another maximum when speaker 1 is at $x = 0.90$ m.
    a. What is the frequency of the sound? Assume $v_{sound} = 340$ m/s.
    b. What is the phase difference between the speakers?
62. A sheet of glass is coated with a 500-nm-thick layer of oil ($n = 1.42$).
    a. For what *visible* wavelengths of light do the reflected waves interfere constructively?
    b. For what *visible* wavelengths of light do the reflected waves interfere destructively?
    c. What is the color of reflected light? What is the color of transmitted light?
63. A jewelry maker has asked your glass studio to produce a sheet of dichroic glass that will appear red for transmitted light and blue for reflected light. You decide that "red light" should be centered at 640 nm and that "blue light" is 480 nm. If you use a MgF$_2$ coating ($n = 1.39$), how thick should the coating be?

64. Example 21.10 showed that a 92-nm-thick coating of MgF$_2$ ($n = 1.39$) on glass acts as an antireflection coating for light with a wavelength of 510 nm. Without the coating, the intensity of reflected light is $I_0 = Ca^2$, where a is the amplitude of the reflected light wave and C is an unknown proportionality constant.
    a. Let I_λ be the intensity of light reflected from the coated glass at wavelength λ. Find an expression for the ratio I_λ/I_0 as a function of the wavelength λ. This ratio is the reflection intensity from the coated glass relative to the reflection intensity from uncoated glass. A ratio less than 1 indicates that the coating is reducing the reflection intensity.
    Hint: The amplitude of the superposition of two waves depends on the phase difference between the waves. Although not entirely accurate, assume that both reflected waves have amplitude a.
    b. Evaluate I_λ/I_0 at $\lambda = 400, 450, 500, 550, 600, 650,$ and 700 nm. This spans the range of visible light.
    c. Draw a graph of I_λ/I_0 versus λ.
65. A manufacturing firm has hired your company, Acoustical Consulting, to help with a problem. Their employees are complaining about the annoying hum from a piece of machinery. Using a frequency meter, you quickly determine that the machine emits a rather loud sound at 1200 Hz. After investigating, you tell the owner that you cannot solve the problem entirely, but you can at least improve the situation by eliminating reflections of this sound from the walls. You propose to do this by installing mesh screens in front of the walls. A portion of the sound will reflect from the mesh; the rest will pass through the mesh and reflect from the wall. How far should the mesh be placed in front of the wall for this scheme to work?
66. A soap bubble is essentially a very thin film of water ($n = 1.33$) surrounded by air. The colors that you see in soap bubbles are produced by interference, much like the colors of dichroic glass.
    a. Derive an expression for the wavelengths λ_C for which constructive interference causes a strong reflection from a soap bubble of thickness d.
    Hint: Think about the reflection phase shifts at both boundaries.
    b. What visible wavelengths of light are strongly reflected from a 390-nm-thick soap bubble? What color would such a soap bubble appear to be?
67. Two radio antennas are separated by 2.0 m. Both broadcast identical 750 MHz waves. If you walk around the antennas in a circle of radius 10 m, how many maxima will you detect?
68. You are standing 2.5 m directly in front of one of the two loudspeakers shown in Figure P21.68. They are 3.0 m apart and both are playing a 686 Hz tone in phase. As you begin to walk directly away from the speaker, at what distances from the speaker do you hear a *minimum* sound intensity? The room temperature is 20°C.

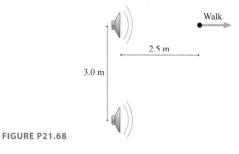

FIGURE P21.68

69. Two loudspeakers 5.0 m apart are playing the same frequency. If you stand 12.0 m in front of the plane of the speakers, centered between them, you hear a sound of maximum intensity. As you walk parallel to the plane of the speakers, staying 12.0 m in front of them, you first hear a minimum of sound intensity when you are directly in front of one of the speakers.
    a. What is the frequency of the sound? Assume a sound speed of 340 m/s.
    b. If you stay 12.0 m in front of one of the speakers, for what other frequencies between 100 Hz and 1000 Hz is there a minimum sound intensity at this point?

70. Two in-phase loudspeakers are located at (x, y) coordinates $(-3.0 \text{ m}, +2.0 \text{ m})$ and $(-3.0 \text{ m}, -2.0 \text{ m})$. They emit identical sound waves with a 2.0 m wavelength and amplitude a. Determine the amplitude of the sound at the five positions on the y-axis $(x = 0)$ with $y = 0.0$ m, 0.5 m, 1.0 m, 1.5 m, and 2.0 m.

71. Your firm has been hired to design a system that allows airplane pilots to make instrument landings in rain or fog. You've decided to place two radio transmitters 50 m apart on either side of the runway. These two transmitters will broadcast the same frequency, but out of phase with each other. This will cause a nodal line to extend straight off the end of the runway (see Figure 21.29b). As long as the airplane's receiver is silent, the pilot knows she's directly in line with the runway. If she drifts to one side or the other, the radio will pick up a signal and sound a warning beep. To have sufficient accuracy, the first intensity maxima need to be 60 m on either side of the nodal line at a distance of 3.0 km. What frequency should you specify for the transmitters?

72. Two radio antennas are 100 m apart along a north-south line. They broadcast identical radio waves at a frequency of 3.0 MHz. Your job is to monitor the signal strength with a handheld receiver. To get to your first measuring point, you walk 800 m east from the midpoint between the antennas, then 600 m north.
    a. What is the phase difference between the waves at this point?
    b. Is the interference at this point maximum constructive, perfect destructive, or somewhere in between? Explain.
    c. If you now begin to walk farther north, does the signal strength increase, decrease, or stay the same? Explain.

73. The three identical loudspeakers in Figure P21.73 play a 170 Hz tone in a room where the speed of sound is 340 m/s. You are standing 4 m in front of the middle speaker. At this point, the amplitude of the wave from each speaker is a.
    a. What is the amplitude at this point?
    b. How far must speaker 2 be moved to the left to produce a maximum amplitude at the point where you are standing?
    c. When the amplitude is maximum, by what factor is the sound intensity greater than the sound intensity from a single speaker?

3.0 m

3.0 m

4.0 m

FIGURE P21.73

74. Piano tuners tune pianos by listening to the beats between the *harmonics* of two different strings. When properly tuned, the note A should have a frequency of 440 Hz and the note E should be at 659 Hz.
    a. What is the frequency difference between the third harmonic of the A and the second harmonic of the E?
    b. A tuner first tunes the A string very precisely by matching it to a 440 Hz tuning fork. She then strikes the A and E strings simultaneously and listens for beats between the harmonics. What beat frequency indicates that the E string is properly tuned?
    c. The tuner starts with the tension in the E string a little low, then tightens it. What is the frequency of the E string when she hears four beats per second?

75. A flutist assembles her flute in a room where the speed of sound is 342 m/s. When she plays the note A, it is in perfect tune with a 440 Hz tuning fork. After a few minutes, the air inside her flute has warmed to where the speed of sound is 346 m/s.
    a. How many beats per second will she hear if she now plays the note A as the tuning fork is sounded?
    b. How far does she need to extend the "tuning joint" of her flute to be in tune with the tuning fork?

76. Two lasers with very nearly the same wavelength can generate a beat frequency if both laser beams illuminate a photodetector with a very fast response. In an experiment, one laser's wavelength has been stabilized at 780.54510 nm. The second laser starts with a longer wavelength that is slowly decreased until the beat frequency between the two lasers is 98.5 MHz. What is the second laser's wavelength?

77. Two loudspeakers emit 400 Hz notes. One speaker sits on the ground. The other speaker is in the back of a pickup truck. You hear eight beats per second as the truck drives away from you. What is the truck's speed?

78. Two loudspeakers face each other from opposite walls of a room . Both are playing exactly the same frequency, thus setting up a standing wave with distance $\lambda/2$ between antinodes. Assume that λ is much less than the room width, so there are many antinodes.
    a. Yvette starts at one speaker and runs toward the other at speed v_Y. As the does so, she hears a loud-soft-loud modulation of the sound intensity. From your perspective, as you sit at rest in the room, Yvette is running through the nodes and antinodes of the standing wave. Find an expression for the number of sound maxima she hears per second.
    b. From Yvette's perspective, the two sound waves are Doppler shifted. They're not the same frequency, so they don't create a standing wave. Instead, she hears a loud-soft-loud modulation of the sound intensity because of beats. Find an expression for the beat frequency that Yvette hears.
    c. Are your answers to parts a and b the same or different? *Should* they be the same or different?

Challenge Problems

79. a. The frequency of a standing wave on a string is f when the string's tension is T. If the tension is changed by the *small* amount ΔT, without changing the length, show that the frequency changes by an amount Δf such that

$$\frac{\Delta f}{f} = \frac{1}{2}\frac{\Delta T}{T}$$

b. Two identical strings vibrate at 500 Hz when stretched with the same tension. What percentage increase in the tension of one of the strings will cause five beats per second when both strings vibrate simultaneously?

80. A 280 Hz sound wave is directed into one end of a trombone slide and a microphone is placed at the other end to record the intensity of sound waves that are transmitted through the tube. The straight sides of the slide are 80 cm in length and 10 cm apart with a semicircular bend at the end. For what slide extensions s will the microphone detect a maximum of sound intensity?

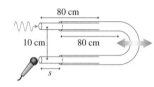

FIGURE CP21.80

81. As the captain of the scientific team sent to Planet Physics, one of your tasks is to measure g. You have a long, thin wire labeled 1.00 g/m and a 1.25 kg weight. You have your accurate space cadet chronometer but, unfortunately, you seem to have forgotten a meter stick. Undeterred, you first find the midpoint of the wire by folding it in half. You then attach one end of the wire to the wall of your laboratory, stretch it horizontally to pass over a pulley at the midpoint of the wire, then tie the 1.25 kg weight to the end hanging over the pulley. By vibrating the wire, and measuring time with your chronometer, you find that the wire's second harmonic frequency is 100 Hz. Next, with the 1.25 kg weight still tied to one end of the wire, you attach the other end to the ceiling to make a pendulum. You find that the pendulum requires 314 s to complete 100 oscillations. Pulling out your trusty calculator, you get to work. What value of g will you report back to headquarters?

82. A 22-cm-long, 1.0-mm-diameter copper wire is joined smoothly to a 60-cm-long, 1.0-mm-diameter aluminum wire. The resulting wire is stretched with 20 N of tension between fixed supports 82 cm apart. The densities of copper and aluminum are 8920 kg/m³ and 2700 kg/m³, respectively.
   a. What is the lowest-frequency standing wave for which there is a node at the junction between the two metals?
   b. At that frequency, how many antinodes are on the aluminum wire?

83. Ultrasound has many medical applications, one of which is to monitor fetal heartbeats by reflecting ultrasound off a fetus in the womb.
   a. Consider an object moving at speed v_o toward an at-rest source that is emitting sound waves of frequency f_0. Show that the reflected wave (i.e., the echo) that returns to the source has a Doppler-shifted frequency

$$f_{echo} = \left(\frac{v + v_o}{v - v_o}\right)f_0$$

where v is the speed of sound in the medium.
   b. Suppose the object's speed is much less than the wave speed: $v_o \ll v$. Then $f_{echo} \approx f_0$, and a microphone that is sensitive to these frequencies will detect a beat frequency if it listens to f_0 and f_{echo} simultaneously. Use the binomial

approximation and other appropriate approximations to show that the beat frequency is $f_{beat} \approx (2v_0/v)f_0$.
   c. The reflection of 2.40 MHz ultrasound waves from the surface of a fetus's beating heart is combined with the 2.40 MHz wave to produce a beat frequency that reaches a maximum of 65 Hz. What is the maximum speed of the surface of the heart? The speed of ultrasound waves within the body is 1540 m/s.
   d. Suppose the surface of the heart moves in simple harmonic motion at 90 beats/min. What is the amplitude in mm of the heartbeat?

84. A water wave is called a *deep-water wave* if the water's depth is more than one-quarter of the wavelength. Unlike the waves we've considered in this chapter, the speed of a deep water wave depends on its wavelength:

$$v = \sqrt{\frac{g\lambda}{2\pi}}$$

Longer wavelengths travel faster. Let's apply this to standing waves. Consider a diving pool that is 5.0 m deep and 10.0 m wide. Standing water waves can set up across the width of the pool. Because water sloshes up and down at the sides of the pool, the boundary conditions require antinodes at $x = 0$ and $x = L$. Thus a standing water wave resembles a standing sound wave in an open-open tube.
   a. What are the wavelengths of the first three standing-wave modes for water in the pool? Do they satisfy the condition for being deep-water waves? Draw a graph of each.
   b. What are the wave speeds for each of these waves?
   c. Derive a general expression for the frequencies f_m of the possible standing waves. Your expression should be in terms of m, g, and L.
   d. What are the oscillation *periods* of the first three standing-wave modes?

85. The broadcast antenna of an AM radio station is located at the edge of town. The station owners would like to beam all of the energy into town and none into the countryside, but a single antenna radiates energy equally in all directions. Figure CP21.85 shows two parallel antennas separated by distance L. Both antennas broadcast a signal at wavelength λ, but antenna 2 can delay its broadcast relative to antenna 1 by a time interval Δt in order to create a phase difference $\Delta\phi_0$ between the sources. Your task is to find values of L and Δt such that the waves interfere constructively on the town side and destructively on the country side.

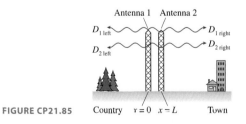

FIGURE CP21.85   Country $x = 0$ $x = L$ Town

Let antenna 1 be at $x = 0$. The wave that travels to the right is $a\sin[2\pi(x/\lambda - t/T)]$. The wave that travels to the left is $a\sin[2\pi(-x/\lambda - t/T)]$. (It must be this, rather than $a\sin[2\pi(x/\lambda + t/T)]$, so that the two waves match at $x = 0$.) Antenna 2 is at $x = L$. It broadcasts waves

$a \sin[2\pi((x - L)/\lambda - t/T) + \phi_{20}]$ to the right and $a \sin[2\pi(-(x - L)/\lambda - t/T) + \phi_{20}]$ to the left.

a. What is the smallest value of L for which you can create perfect constructive interference on the town side and perfect destructive interference on the country side? Your answer will be a multiple or fraction of the wavelength λ.

b. What phase constant ϕ_{20} of antenna 2 is needed?

c. What fraction of the oscillation period T must Δt be to produce the proper value of ϕ_{20}?

d. Evaluate both L and Δt for the realistic AM radio frequency of 1000 KHz.

Comment: This is a simple example of what is called a *phased array,* where phase differences between identical emitters are used to "steer" the radiation in a particular direction. Phased arrays are widely used in radar technology.

STOP TO THINK ANSWERS

Stop to Think 21.1: c. The figure shows the two waves at $t = 6$ s and their superposition. The superposition is the *point-by-point* addition of the displacements of the two individual waves.

Stop to Think 21.2: a. The allowed standing-wave frequencies are $f_m = m(v/2L)$, so the mode number of a standing wave of frequency f is $m = 2Lf/v$. Quadrupling T_s increases the wave speed v by a factor of two. The initial mode number was 2, so the new mode number is 1.

Stop to Think 21.3: b. 300 Hz and 400 Hz are allowed standing waves, but they are not f_1 and f_2 because 400 Hz $\neq$ 2 $\times$ 300 Hz. Because there's a 100 Hz difference between them, these must be $f_3 = 3 \times 100$ Hz and $f_4 = 4 \times 100$ Hz, with a fundamental frequency $f_1 = 100$ Hz. Thus the second harmonic is $f_2 = 2 \times 100$ Hz $= 200$ Hz.

Stop to Think 21.4: c. Shifting the top wave 0.5 m to the left aligns crest with crest and trough with trough.

Stop to Think 21.5: c. $r_1 = 0.5\lambda$ and $r_2 = 3.0\lambda$, so $\Delta r = 2.5\lambda$. This is the condition for perfect destructive interference.

Stop to Think 21.6: Maximum constructive. The path-length difference is $\Delta r = 1.0$ m $= \lambda$. For identical sources, interference is constructive when Δr is an integer multiple of λ.

Stop to Think 21.7: f. The beat frequency is the difference between the two frequencies.

22 Wave Optics

► Looking Ahead

The goal of Chapter 22 is to understand and apply the wave model of light. In this chapter you will learn to:

- Use the wave model of light.
- Recognize experimental evidence for the wave nature of light.
- Calculate the interference pattern of double slits and diffraction gratings.
- Understand how light diffracts through single slits and circular apertures.
- Understand how interferometers control the interference of light.

◄ Looking Back

Wave optics depends on the basic properties of waves that have been developed in Chapters 20 and 21. Please review:

- Sections 20.4–20.6 Wave fronts, phase, and intensity as they pertain to light waves.
- Section 21.7 Interference in two and three dimensions.

You've probably noticed the rainbow splash of colors when a bright light reflects from the surface of a compact disk. You may be surprised to learn that the colors from a CD are closely related to the iridescence of bird feathers and to the technology underlying supermarket checkout scanners, holograms, and optical computers. All of these, in one way or another, depend on the interference of light waves. In some, such as the CD, the interference is incidental to the intended use of the product. Others, such as holograms, have been designed to make precise use of the wave-like properties of light.

The study of light is called **optics,** and this is the first of three chapters to explore optics and the nature of light. Light is an elusive topic. You will find, perhaps surprisingly, that there is no simple description of light. Light behaves quite differently in different situations, and we will ultimately need three different *models* of light to capture this behavior. We begin, in this chapter, with situations in which light acts as a wave. The groundwork for *wave optics* has been laid in Chapters 20 and 21, and we will now apply those ideas to light waves.

Although light is an electromagnetic wave, this chapter depends on nothing more than the "waviness" of light waves. You can study this chapter either before or after your study of electricity and magnetism in Part VI.

22.1 Light and Optics

What is light? The first Greek scientists and philosophers did not make a distinction between light and vision. Light, to them, was not something that existed apart from seeing. But gradually there arose a view that light actually "exists," that light is some sort of physical entity that is present regardless of whether or not someone is looking. But if light is a physical entity, what is it? What are its characteristics? Is it a wave, similar to sound? Or is light a collection of small particles that blows by like the wind?

Newton, in addition to doing pioneering work in mathematics and mechanics in the 1660s, was also one of the early investigators of the nature of light. Newton knew that a water wave, after passing through an opening, *spreads out* to fill the space behind the opening. You can see this in Figure 22.1a, where plane waves, approaching from the left, spread out in circular arcs after passing through a hole in a barrier. This inexorable spreading of waves is the phenomenon called **diffraction.** Diffraction is a sure sign that whatever is passing through the hole is a wave.

In contrast, Figure 22.1b shows that sunlight passing through a door makes a sharp-edged shadow as it falls upon the floor. We don't see sunlight light spreading out after it passes through a hole. Instead, this behavior is exactly what you would expect if light consists of particles traveling in straight lines. Some particles would pass through the window to make a bright area on the wall, others would be blocked and cause a well-defined shadow. This reasoning led Newton to the conclusion that light consists of very small, light, fast particles that he called *corpuscles.*

Newton was vigorously opposed by Robert Hooke (of Hooke's law) and the Dutch scientist Christian Huygens, who argued that light was some sort of wave. Although the debate was lively, and sometimes acrimonious, Newton eventually prevailed. The belief that light consists of corpuscles was not seriously questioned for more than a hundred years after Newton's death.

The situation changed dramatically in 1801, when the English scientist Thomas Young announced that he had produced *interference* between two waves of light. Young's experiment, which we will analyze in the next section, was a painstakingly difficult experiment with the technology of his era. Nonetheless, Young's experiment quickly settled the debate in favor of a wave theory of light because interference is a distinctly wave-like phenomenon.

But if light is a wave, what is it that is waving? This was the question that Young posed to the 19th century. It was ultimately established, through theoretical and experimental efforts by numerous scientists, that light is an *electromagnetic wave,* an oscillation of the electromagnetic field that requires no material medium in which to travel. Further, as we have already seen, visible light is just one small slice out of a vastly broader *electromagnetic spectrum.*

That light is a wave, an electromagnetic wave, seemed well established by about 1880. But this satisfying conclusion was undermined within the next 25 years. A new discovery, called the photoelectric effect, seemed to be inconsistent with the theory of electromagnetic waves. In 1905, an unknown young physicist named Albert Einstein was able to explain the photoelectric effect by treating light as a novel type of wave having certain particle-like characteristics. These wave-like particles of light soon came to be known as *photons.*

Einstein's introduction of the concept of the photon can now be seen as the end of *classical physics* and the beginning of a new era called *quantum physics.* Equally important, Einstein's theory marked yet another shift in our age-old effort to understand light.

(a) Plane waves approach from the left.

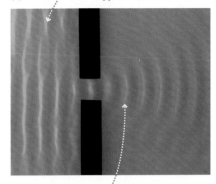

Circular waves spread out on the right.

(b)

A beam of sunlight has a sharp edge.

FIGURE 22.1 Water waves spread out behind a small hole in a barrier, but light passing through a doorway makes a sharp-edged shadow.

Models of Light

Light is a real physical entity, but the nature of light is elusive. Light is the chameleon of the physical world. Under some circumstances, light acts like particles traveling in straight lines. But change the circumstances, and light shows the same kinds of wave-like behavior as sound waves or water waves. Change the circumstances yet again, and light exhibits behavior that is neither wave-like nor particle-like but has characteristics of both.

Rather than an all-encompassing "theory of light," it will be better to develop several **models of light.** Each model successfully explains the behavior of light within a certain domain—that is, within a certain range of physical situations. Our task will be twofold:

1. To develop clear and distinct models of light.
2. To learn the conditions and circumstances for which each model is valid.

The second task is especially important.

We'll begin with a brief summary of all three models, so that you will have a road map of where we're headed. Each of these models will be developed in the coming chapters.

The wave model: The wave model of light is the most widely applicable model, responsible for the widely known "fact" that light is a wave. It is certainly true that, under many circumstances, light exhibits the same behavior as sound or water waves. Lasers and electro-optical devices, critical technologies of the 21st century, are best understood in terms of the wave model of light. Some aspects of the wave model of light were introduced in Chapters 20 and 21, and the wave model is the primary focus of this chapter. The study of light as a wave is called **wave optics.**

The ray model: An equally well-known "fact" is that light travels in a straight line. These straight-line paths are called *light rays.* In Newton's view, light rays are the trajectories of particle-like corpuscles of light. The properties of prisms, mirrors, lenses, and optical instruments such as telescopes and microscopes are best understood in terms of light rays. Unfortunately, it's difficult to reconcile the statement "light travels in a straight line" with the statement "light is a wave." For the most part, waves and rays are mutually exclusive models of light. One of our most important tasks will be to learn when each model is appropriate. The ray model of light, the basis of **ray optics,** is the subject of the next chapter.

The photon model: Modern technology is increasingly reliant on quantum physics. In the quantum world, light behaves like neither a wave nor a particle. Instead, light consists of *photons* that have both wave-like and particle-like properties. Photons are the *quanta* of light. Much of the quantum theory of light is beyond the scope of this textbook, but we will take a peek at the important ideas in Chapter 24 and again in Part VII.

22.2 The Interference of Light

Suppose that Newton had seen the experiment depicted in Figure 22.2. Here light passes through a "window" that is only 0.1 mm wide, about twice the width of a human hair. The photograph shows how the light appears on a viewing screen 2 m behind the aperture. If light consists of corpuscles traveling in straight lines, as Newton thought, we should see a narrow strip of light, about 0.1 mm wide, with dark shadows on cither side. Instead, we see a band of light extending over about 2.5 cm, a distance much wider than the aperture, with dimmer patches of light extending even farther on either side.

If you compare Figure 22.2 to the water wave of Figure 22.1, you see that *the light is spreading out* behind the 0.1-mm-wide hole. The light is exhibiting diffraction, the sure signature of waviness. The diffraction of light wasn't known in

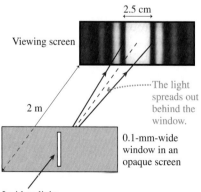

FIGURE 22.2 Light, just like a water wave, does spread out behind a hole *if* the hole is sufficiently small.

Newton's time because, as we shall see, diffraction is apparent only for very narrow apertures, typically less than 0.5 mm in width. Such apertures are hard to make, and, because they are so narrow, the light on the viewing screen is extremely dim unless you have a very bright source of light. Consequently, we can surmise that Newton would have reached a very different conclusion had he been able to perform this experiment.

We will look at diffraction in more detail later in the chapter. For now, we merely need the *observation* that light does, indeed, spread out behind a hole that is sufficiently small.

Young's Double-Slit Experiment

Rather than one small hole, suppose we use two. Figure 22.3a shows an experiment in which a laser beam is aimed at an opaque screen containing two long, narrow slits that are very close together. This pair of slits is called a **double slit,** and in a typical experiment they are ≈0.1 mm wide and spaced ≈0.5 mm apart. We will assume that the laser beam illuminates both slits equally, and any light passing through the slits impinges on a viewing screen. This is the essence of Young's experiment of 1801, although he used sunlight rather than a laser.

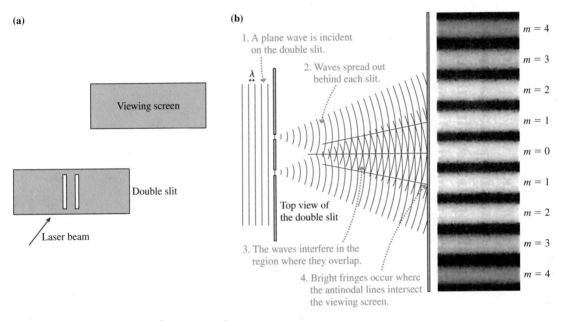

FIGURE 22.3 A double-slit interference experiment.

What should we expect to see on the screen? Figure 22.3b is a view from above the experiment, looking down on the top ends of the slits and the top edge of the viewing screen. Because the slits are very narrow, **light spreads out behind each slit** as it did in Figure 22.2, and these two spreading waves overlap in the region between the slits and the screen.

The primary conclusion of Chapter 21 was that two overlapped waves of equal wavelength produce interference. In fact, Figure 22.3b is equivalent to the waves emitted by two loudspeakers, a situation we analyzed in Section 21.7. (It is very useful to compare Figure 22.3b with Figures 21.27 and 21.29a.) Nothing in that analysis depended on what type of wave it was, so the conclusions apply equally well to two overlapped light waves. If light really is a wave, we should see interference between the two light waves over the small region, typically 2 or 3 cm wide, where they overlap on the viewing screen.

The photograph in Figure 22.3b shows how the screen looks. As expected, the light is intense at points where an antinodal line intersects the screen. There is no light at all at points where a nodal line intersects the screen. These alternating bright and dark bands of light, due to constructive and destructive interference, are called **interference fringes.** The fringes are numbered $m = 0, 1, 2, 3, \ldots$, going outward from the center. The brightest fringe, at the midpoint of the viewing screen, with $m = 0$, is called the **central maximum.**

> **STOP TO THINK 22.1** Suppose the viewing screen in Figure 22.3 is moved closer to the double slit. What happens to the interference fringes?
>
> a. They get brighter but otherwise do not change.
> b. They get brighter and closer together.
> c. They get brighter and farther apart.
> d. They get out of focus.
> e. They fade out and disappear.

Analyzing Double-Slit Interference

16.1–16.3 Activ Physics ONLINE

Figure 22.3 showed qualitatively that interference is produced behind a double slit by the overlap of the light waves spreading out behind each opening. Now let's analyze the experiment more carefully. Figure 22.4 shows the geometry of a double-slit experiment in which the spacing between the two slits is d and the distance to the viewing screen is L. We will assume that L is *very* much larger than d.

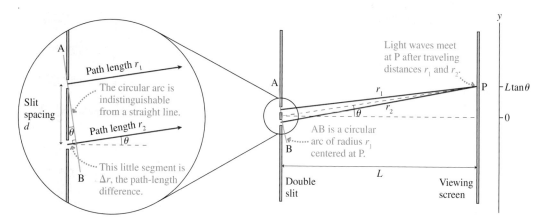

FIGURE 22.4 Geometry of the double-slit experiment.

Our goal, as it was with sound waves in Chapter 21, is to determine if the interference at a particular point is constructive, destructive, or in between. Let P be a point on the screen that is distance r_1 from one slit and r_2 from the other. We can specify point P either by the distance y from the midpoint on the viewing screen or by the angle θ from the midpoint between the slits. Angle θ and distance y are related by

$$y = L\tan\theta \qquad (22.1)$$

Figure 22.3b showed that both slits are illuminated by the *same* wave front from the laser. Consequently, the slits act as sources of identical waves ($\Delta\phi_0 = 0$). You learned in Chapter 21 that constructive interference between the waves from identical sources occurs at points for which the path-length difference $\Delta r = r_2 - r_1$ is an integer number of wavelengths:

$$\Delta r = m\lambda \qquad m = 0, 1, 2, 3, \ldots \qquad \text{(constructive interference)} \qquad (22.2)$$

Thus the interference at point P is constructive, producing a bright fringe, if $\Delta r = m\lambda$ at that point.

The midpoint on the viewing screen at $y = 0$ is equally distant from both slits ($\Delta r = 0$). Constructive interference at the midpoint produces the bright fringe identified as the central maximum in Figure 21.3b. The path-length difference increases as you move away from the center of the screen, and the $m = 1$ fringes occur at the positions where $\Delta r = 1\lambda$. That is, one wave has traveled exactly one wavelength farther than the other. In general, **the mth bright fringe occurs where one wave has traveled m wavelengths farther than the other and thus $\Delta r = m\lambda$.**

We need to find the specific positions on the screen where $\Delta r = m\lambda$. To do so, draw a circular arc of radius r_1 centered at point P. This arc, labeled AB in Figure 22.4, divides the path from the lower slit into a long segment of length r_1 and a much shorter segment of length $\Delta r = r_2 - r_1$.

Because L is very large in comparison to the slit spacing d, the two paths r_1 and r_2 are essentially parallel and the arc AB is indistinguishable from a straight line forming one side of a right triangle. This is shown in the magnified portion of Figure 22.4, where you can see that

$$\Delta r = d\sin\theta \tag{22.3}$$

Bright fringes (constructive interference) occur at angles θ_m such that

$$\Delta r = d\sin\theta_m = m\lambda \qquad m = 0, 1, 2, 3, \ldots \tag{22.4}$$

We have added the subscript m to denote that θ_m is the angle of the mth bright fringe, starting with $m = 0$ at the center.

In practice, the angle θ in a double-slit experiment is a very small angle ($<1°$) We can use the small-angle approximation $\sin\theta \approx \theta$, where θ must be in radians, to write Equation 22.4 as

$$\theta_m = m\frac{\lambda}{d} \qquad m = 0, 1, 2, 3, \ldots \qquad \text{(angles of bright fringes)} \tag{22.5}$$

This gives the angular positions *in radians* of the bright fringes in the interference pattern.

It is usually more convenient to measure the *position* of the mth bright fringe, as measured from the center of the viewing screen. Using the small-angle approximation once again, this time in the form $\tan\theta \approx \theta$, we can substitute θ_m from Equation 22.5 for $\tan\theta_m$ in Equation 22.1 to find that the mth bright fringe occurs at position

$$y_m = \frac{m\lambda L}{d} \qquad m = 0, 1, 2, 3, \ldots \qquad \text{(positions of bright fringes)} \tag{22.6}$$

The interference pattern is symmetrical, so there is an mth bright fringe at the same distance on both sides of the center. You can see this in Figure 22.3b. As we've already noted, **the $m = 1$ fringes occur at points on the screen where the light from one slit travels exactly one wavelength farther than the light from the other slit.**

NOTE ▶ Equations 22.5 and 22.6 do *not* apply to the interference of sound waves from two loudspeakers. The approximations we've used (small angles, $L \gg d$) are usually not valid for the much longer wavelengths of sound waves. ◀

Equation 22.6 predicts that **the interference pattern is a series of equally spaced bright lines** on the screen, exactly as shown in Figure 22.3b. How do we know the fringes are equally spaced? The **fringe spacing** between the m fringe and the $m + 1$ fringe is

$$\Delta y = y_{m+1} - y_m = \frac{(m + 1)\lambda L}{d} - \frac{m\lambda L}{d} = \frac{\lambda L}{d} \tag{22.7}$$

Because Δy is independent of m, *any* two bright fringes have the same spacing.

The dark fringes in the photograph are bands of destructive interference. You learned in Chapter 21 that destructive interference occurs at positions where the path-length difference of the waves is a half-integer number of wavelengths:

$$\Delta r = \left(m + \frac{1}{2}\right)\lambda \qquad m = 0, 1, 2, 3, \dots \tag{22.8}$$
(destructive interference)

We can use Equation 22.4 for Δr and the small-angle approximation to find that the dark fringes are located at positions

$$y'_m = \left(m + \frac{1}{2}\right)\frac{\lambda L}{d} \qquad m = 0, 1, 2, 3, \dots$$
(positions of dark fringes)
$$\tag{22.9}$$

We have used y'_m, with a prime, to distinguish the location of the mth minimum from the mth maximum at y_m. You can see from Equation 22.9 that **the dark fringes are located exactly halfway between the bright fringes.**

EXAMPLE 22.1 Double-slit interference of a laser beam
Light from a helium-neon laser ($\lambda = 633$ nm) illuminates two slits spaced 0.40 mm apart. A viewing screen is 2.0 m behind the slits. What are the distances between the two $m = 2$ bright fringes and between the two $m = 2$ dark fringes?

MODEL Two closely spaced slits produce a double-slit interference pattern.

VISUALIZE The interference pattern looks like the photograph of Figure 22.3b. It is symmetrical, with $m = 2$ bright fringes at equal distances on both sides of the central maximum.

SOLVE The $m = 2$ bright fringe is located at position

$$y_m = \frac{m\lambda L}{d} = \frac{2(633 \times 10^{-9}\ \text{m})(2.0\ \text{m})}{4.0 \times 10^{-4}\ \text{m}}$$

$$= 6.3 \times 10^{-3}\ \text{m} = 6.3\ \text{mm}$$

Each of the $m = 2$ fringes is 6.3 mm from the central maximum, hence the distance between the two $m = 2$ bright fringes is 12.6 mm. The $m = 2$ dark fringe is located at

$$y'_m = \left(m + \frac{1}{2}\right)\frac{\lambda L}{d} = 7.9\ \text{mm}$$

Thus the distance between the two $m = 2$ dark fringes is 15.8 mm.

ASSESS As the fringes are counted outward from the center, the $m = 2$ bright fringe occurs *before* the $m = 2$ dark fringe.

EXAMPLE 22.2 Measuring the wavelength of light
A double-slit interference pattern is observed on a screen 1.0 m behind two slits spaced 0.30 mm apart. Ten bright fringes span a distance of 1.65 cm. What is the wavelength of the light?

MODEL It is not always obvious which fringe is the central maximum. Slight imperfections in the slits can make the interference fringe pattern less than ideal. However, you do not need to identify the $m = 0$ fringe because you can make use of the

fact that the fringe spacing Δy is uniform. Ten bright fringes have *nine* spaces between them (not ten—be careful!).

VISUALIZE The interference pattern looks like the photograph of Figure 22.3b.

SOLVE The fringe spacing is

$$\Delta y = \frac{1.65\ \text{cm}}{9} = 1.833 \times 10^{-3}\ \text{m}$$

Using this fringe spacing in Equation 22.7, we find that the wavelength is

$$\lambda = \frac{d}{L}\Delta y = 5.50 \times 10^{-7}\ \text{m} = 550\ \text{nm}$$

It is customary to express the wavelengths of light in nanometers. Be sure to do this as you solve problems.

ASSESS Young's double-slit experiment not only demonstrated that light is a wave, it provided a means for measuring the wavelength. You learned in Chapter 20 that the wavelengths of visible light span the range 400–700 nm. These lengths are smaller than we can easily comprehend. A wavelength of 550 nm, which is in the middle of the visible spectrum, is only about 1% of the diameter of a human hair.

STOP TO THINK 22.2   Light of wavelength λ_1 illuminates a double slit, and interference fringes are observed on a screen behind the slits. When the wavelength is changed to λ_2, the fringes get closer together. Is λ_2 larger or smaller than λ_1?

Intensity of the Double-Slit Interference Pattern

Equations 22.6 and 22.9 locate the positions of maximum and zero intensity. To complete our analysis we need to calculate the light *intensity* at every point on the screen. All the tools we need to do this calculation were developed in Chapters 20 and 21.

You learned in Chapter 20 that the wave intensity I is proportional to the square of the wave's amplitude. The light spreading out behind a *single* slit produces the wide band of light that you saw in Figure 22.2. The intensity in this band of light is $I_1 = Ca^2$, where a is the light-wave amplitude at the screen due to *one* wave and C is a proportionality constant.

If there were no interference, the light intensity due to two slits would be twice the intensity of one slit: $I_2 = 2I_1 = 2Ca^2$. In other words, two slits would cause the broad band of light on the screen to be twice as bright. But that's not what happens. Instead, the superposition of the two light waves creates bright and dark interference fringes.

We found in Chapter 21 (Equation 21.36) that the net amplitude of two superimposed waves is

$$A = \left| 2a\cos\left(\frac{\Delta\phi}{2}\right) \right| \tag{22.10}$$

where a is the amplitude of each individual wave. Because the sources are in phase, the phase difference $\Delta\phi$ at the point where the two waves are combined is $\Delta\phi = 2\pi(\Delta r/\lambda)$. Using Equation 22.3 for the path-length difference Δr, along with the small-angle approximation and Equation 22.2 for y gives us the phase difference at position y on the screen:

$$\Delta\phi = 2\pi\frac{\Delta r}{\lambda} = 2\pi\frac{d\sin\theta}{\lambda} \approx 2\pi\frac{d\tan\theta}{\lambda} = \frac{2\pi d}{\lambda L}y \tag{22.11}$$

Substituting Equation 22.11 into Equation 22.10, we find the wave amplitude at position y to be

$$A = \left| 2a\cos\left(\frac{\pi d}{\lambda L}y\right) \right| \tag{22.12}$$

Consequently, the light intensity at position y on the screen is

$$I = CA^2 = 4Ca^2\cos^2\left(\frac{\pi d}{\lambda L}y\right) \tag{22.13}$$

But Ca^2 is I_1, the light intensity of a single slit. Thus the intensity of the double-slit interference pattern at position y is

$$I_{\text{double}} = 4I_1 \cos^2\left(\frac{\pi d}{\lambda L} y\right) \qquad (22.14)$$

Figure 22.5a is a graph of the double-slit intensity versus position y. Notice the unusual orientation of the graph, with the intensity increasing toward the *left* so that the y-axis can match the experimental layout. You can see that the intensity oscillates between dark fringes ($I_{\text{double}} = 0$) and bright fringes ($I_{\text{double}} = 4I_1$). The maxima occur at points where $y_m = m\lambda L/d$. This is what we found earlier for the positions of the bright fringes, so Equation 22.14 is consistent with our initial analysis.

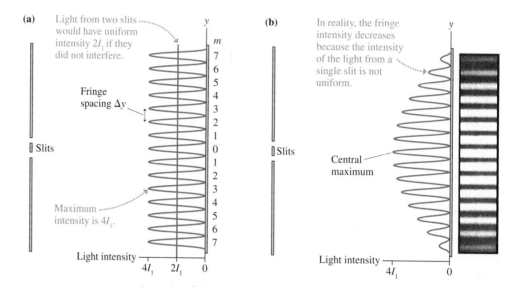

FIGURE 22.5 Intensity of the interference fringes in a double-slit experiment.

One curious feature is that the light intensity at the maxima is $I = 4I_1$, four times the intensity of the light from each slit alone. You might think that two slits would make the light twice as intense as one slit, but interference leads to a different result. Mathematically, two slits make the *amplitude* twice as big at points of constructive interference ($A = 2a$), so the intensity increases by a factor $2^2 = 4$. Physically, this is conservation of energy. The line labeled $2I_1$ in Figure 22.5a is the uniform intensity that two slits would produce *if* the waves did not interfere. Interference does not change the amount of light energy coming through the two slits, but it does redistribute the light energy on the viewing screen. You can see that the *average* intensity of the oscillating curve is $2I_1$, but the intensity of the bright fringes gets pushed up from $2I_1$ to $4I_1$ in order for the intensity of the dark fringes to drop from $2I_1$ to 0.

There is still one problem. Equation 22.14 predicts that all interference fringes are equally bright, but you saw in Figure 22.3b that the fringes decrease in brightness as you move away from the center. The erroneous prediction stems from our assumption that the amplitude a of the wave from each slit is constant across the screen. But this isn't really true. A more detailed calculation, in which the amplitude gradually decreases as you move away from the center, finds that Equation 22.14 is correct if I_1 slowly decreases as y increases.

Figure 22.5b summarizes this analysis by graphing the light intensity (Equation 22.14) with I_1 slowly decreasing as y increases. Comparing this graph to the photograph, you can see that the wave model of light has provided an excellent description of Young's double-slit interference experiment.

22.3 The Diffraction Grating

Suppose we were to replace the double slit with an opaque screen that has N closely spaced slits. When illuminated from one side, each of these slits becomes the source of a light wave that diffracts, or spreads out, behind the slit. Such a multi-slit device is called a **diffraction grating.** The light intensity pattern on a screen behind a diffraction grating is due to the interference of N overlapped waves.

Figure 22.6 shows a diffraction grating in which N slits are equally spaced a distance d apart. This is a top view of the grating, as we look down on the experiment, and the slits extend above and below the page. Only 10 slits are shown here, but a practical grating will have hundreds or even thousands of slits. Suppose a plane wave of wavelength λ approaches from the left. The crest of a plane wave arrives *simultaneously* at each of the slits, causing the wave emerging from each slit to be in phase with the wave emerging from every other slit. Each of these emerging waves spreads out, just like the light wave in Figure 22.2, and after a short distance they all overlap with each other and interfere.

We want to know how the interference pattern will appear on a screen behind the grating. The light wave at the screen is the superposition of N waves, from N slits, as they spread and overlap. As we did with the double slit, we'll assume that the distance L to the screen is very large in comparison with the slit spacing d, hence the path followed by the light from one slit to a point on the screen is *very nearly* parallel to the path followed by the light from neighboring slits. The paths cannot be perfectly parallel, of course, or they would never meet to interfere, but the slight deviation from perfect parallelism is too small to notice. You can see in Figure 22.6 that the wave from one slit travels distance $\Delta r = d\sin\theta$ more than the wave from the slit above it and $\Delta r = d\sin\theta$ less than the wave below it. This is the same reasoning we used in Figure 22.4 to analyze the double-slit experiment.

Figure 22.6 was a magnified view of the slits. Figure 22.7 steps back to where we can see the viewing screen. If the angle θ is such that $\Delta r = d\sin\theta = m\lambda$, where m is an integer, then the light wave arriving at the screen from one slit will be *exactly in phase* with the light waves arriving from the two slits next to it. But each of those waves is in phase with waves from the slits next to them, and so on until we reach the end of the grating. In other words, N **light waves, from N different slits, will *all* be in phase with each other when they arrive at a point on the screen at angle θ_m such that**

$$d\sin\theta_m = m\lambda \qquad m = 0, 1, 2, 3, \ldots \qquad (22.15)$$

The screen will have bright constructive-interference fringes at the values of θ_m given by Equation 22.15. When this happens, we say that the light is "diffracted at angle θ_m." Because it's usually easier to measure distances rather than angles, the distance y_m from the center to the mth maximum is

$$y_m = L\tan\theta_m \qquad \text{(positions of bright fringes)} \qquad (22.16)$$

The integer m is called the **order** of the diffraction. For example, light diffracted at $\theta_2 = 60°$ would be the second-order diffraction. Practical gratings, with very small values for d, display only a few orders. Because d is usually very small, it is customary to characterize a grating by the number of *lines per millimeter.* Here "line" is synonymous with "slit," so the number of lines per millimeter is simply the inverse of the slit spacing d in millimeters.

> **NOTE ▶** The condition for constructive interference in a grating of N slits is identical to Equation 22.4 for just two slits. Equation 22.15 is simply the requirement that the path-length difference between adjacent slits, be they two

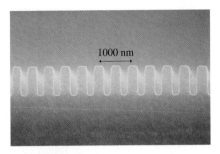

A microscopic side-on look at a diffraction grating.

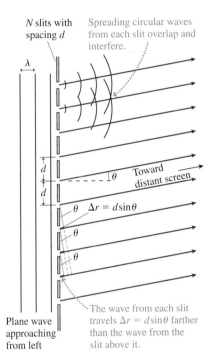

FIGURE 22.6 Top view of a diffraction grating with $N = 10$ slits.

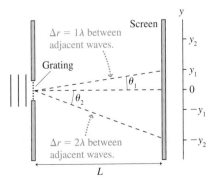

FIGURE 22.7 Angles of constructive interference.

16.4, 16.5   Activ ONLINE Physics

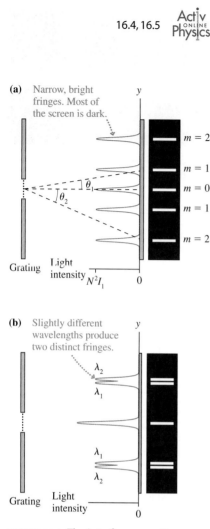

(a) Narrow, bright fringes. Most of the screen is dark.

$m = 2$
$m = 1$
$m = 0$
$m = 1$
$m = 2$

Grating   Light intensity $N^2 I_1$ 0

(b) Slightly different wavelengths produce two distinct fringes.

λ_2
λ_1

λ_1
λ_2

Grating   Light intensity   0

FIGURE 22.8 The interference pattern behind a diffraction grating.

or N, is $m\lambda$. But unlike the angles in double-slit interference, the angles of constructive interference from a diffraction grating are generally *not* small angles. The reason is that the slit spacing d in a diffraction grating is so small that λ/d is not a small number. Thus you *cannot* use the small-angle approximation to simplify Equations 22.15 and 22.16. ◄

The wave amplitude at the points of constructive interference is Na because N waves of amplitude a combine in phase. Because the intensity depends on the square of the amplitude, the intensities of the bright fringes of a diffraction grating are

$$I_{max} = N^2 I_1 \qquad (22.17)$$

where, as before, I_1 is the intensity of the wave from a single slit. Equation 22.17 is consistent with our prior conclusion that the intensity of a bright fringe in a double-slit interference experiment is four times the intensity of the light from each slit alone. You can see that the fringe intensities increase rapidly as the number of slits increases.

Not only do the fringes get brighter as N increases, they also get narrower. This is again a matter of conservation of energy. If the light waves did not interfere, the intensity from N slits would be NI_1. Interference increases the intensity of the bright fringes by an extra factor of N, so to conserve energy the width of the bright fringes must be proportional to $1/N$. For a realistic diffraction grating, with $N > 100$, the interference pattern consists of a small number of *very* bright and *very* narrow fringes while most of the screen remains dark. Figure 22.8a shows the interference pattern behind a diffraction grating both graphically and with a simulation of the viewing screen. A comparison with Figure 22.5b shows that the bright fringes of a diffraction grating are much sharper and more distinct than the fringes of a double slit.

Because the bright fringes are so distinct, diffraction gratings are used for measuring the wavelengths of light. Suppose the incident light consists of two slightly different wavelengths. Each wavelength will be diffracted at a slightly different angle and, if N is sufficiently large, we'll see two distinct fringes on the screen. Figure 22.8b illustrates this idea. By contrast, the fringes in a double-slit experiment are so broad that it would not be possible to distinguish the fringes of one wavelength from those of the other.

EXAMPLE 22.3 Measuring wavelengths emitted by sodium atoms

Light from a sodium lamp passes through a diffraction grating having 1000 slits per millimeter. The interference pattern is viewed on a screen 1.000 m behind the grating. Two bright yellow fringes are visible 72.88 cm and 73.00 cm from the central maximum. What are the wavelengths of these two fringes?

VISUALIZE This is the situation shown in Figure 22.8b. The two fringes are very close together, so we expect the wavelengths to be only slightly different. No other yellow fringes are mentioned, so we will assume these two fringes are the first-order diffraction ($m = 1$).

SOLVE The distance y_m of a bright fringe from the central maximum is related to the diffraction angle by $y_m = L\tan\theta_m$. Thus the diffraction angles of these two fringes are

$$\theta_1 = \tan^{-1}\left(\frac{y_1}{L}\right) = \begin{cases} 36.08° & \text{fringe at 72.88 cm} \\ 36.13° & \text{fringe at 73.00 cm} \end{cases}$$

These angles must satisfy the interference condition $d\sin\theta_1 = \lambda$, so the wavelengths are

$$\lambda = d\sin\theta_1$$

What is d? If a 1 mm length of the grating has 1000 slits, then the spacing from one slit to the next must be 1/1000 mm, or $d = 1.00 \times 10^{-6}$ m. Thus the wavelengths creating the two bright fringes are

$$\lambda = d\sin\theta_1 = \begin{cases} 589.0 \text{ nm} & \text{fringe at 72.88 cm} \\ 589.6 \text{ nm} & \text{fringe at 73.00 cm} \end{cases}$$

ASSESS We had data accurate to four significant figures, and all four were necessary to distinguish the two wavelengths.

The science of measuring the wavelengths of atomic and molecular emissions is called **spectroscopy.** The two sodium wavelengths in this example are called the *sodium doublet,* a name given to two closely spaced wavelengths emitted by the atoms of one element. This doublet is an identifying characteristic of sodium. Because no other element emits these two wavelengths, the doublet can be used to identify the presence of sodium in a sample of unknown composition, even if sodium is only a very minor constituent. This procedure is called *spectral analysis.*

Reflection Gratings

We have analyzed what is called a *transmission grating,* with many parallel slits. In practice, most diffraction gratings are manufactured as *reflection gratings.* The simplest reflection grating, shown in Figure 22.9a, is a mirror with hundreds or thousands of narrow, parallel grooves cut into the surface. The grooves divide the surface into many parallel reflective stripes, each of which, when illuminated, becomes the source of a spreading wave. Thus an incident light wave is divided into N overlapped waves. The interference pattern is exactly the same as the interference pattern of light transmitted through N parallel slits.

Reflection gratings can also be created by carving or molding a series of narrow, parallel ridges into a reflective surface, as illustrated in Figure 22.9b. The *iridescence* of some bird feathers and insect shells arises from biological structures that have small, parallel ridges. These ridges form a reflection grating that diffracts the different wavelengths of sunlight at different angles. Multicolored hues of iridescence appear as the angle between the grating and your eye changes.

The rainbow of colors reflected from the surface of a CD is a similar display of interference. The surface of a CD is smooth plastic with a mirror-like reflective coating. Millions and millions of microscopic holes, each about 1 μm in diameter, are "burned" into the surface with a laser. The presence or absence of a hole at a particular location on the disk is interpreted as the 0 or 1 of digitally encoded information. But from an optical perspective, the array of holes in a shiny surface is a two-dimensional version of the reflection grating shown in Figure 22.9a. Less precise plastic reflection gratings can be manufactured at very low cost simply by stamping holes or grooves into a reflective surface, and these are widely sold as toys and novelty items. Rainbows of color are seen as each wavelength of white light is diffracted at a unique angle.

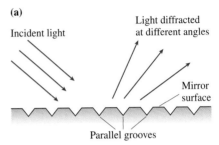

(a)

Incident light

Light diffracted at different angles

Mirror surface

Parallel grooves

A reflection grating can be made by cutting parallel grooves in a mirror surface. These can be very precise, for scientific use, or mass produced in plastic.

(b)

Few μm

Naturally occurring microscopic ridges are present in some bird feathers and insect shells. These cause iridescence when white light reflects off them.

FIGURE 22.9 Reflection gratings.

> **STOP TO THINK 22.3**   White light passes through a diffraction grating and forms rainbow patterns on a screen behind the grating. For each rainbow,
>
> a. the red side is on the right, the violet side on the left.
> b. the red side is on the left, the violet side on the right.
> c. the red side is closest to the center of the screen, the violet side is farthest from the center.
> d. the red side is farthest from the center of the screen, the violet side is closest to the center.

22.4 Single-Slit Diffraction

We opened this chapter with a photograph of a water wave passing through a hole in a barrier, then spreading out on the other side. You then saw a photograph showing that light, after passing through a very narrow slit, also spreads out on the other side. This phenomenon is called *diffraction.* We're now ready to look at the details of diffraction.

Activ
Physics   16.6
ONLINE

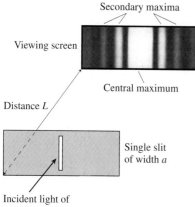

FIGURE 22.10 A single-slit diffraction experiment.

Figure 22.10 shows the experimental arrangement for observing the diffraction of light through a narrow slit of width a. Diffraction through a tall, narrow slit is known as **single-slit diffraction.** A viewing screen is placed distance L behind the slit, and we will assume that $L \gg a$. The light pattern on the viewing screen consists of a *central maximum* flanked by a series of weaker **secondary maxima** and dark fringes. Notice that the central maximum is significantly broader than the secondary maxima. It is also significantly brighter than the secondary maxima, although that is hard to tell here because this photograph has been overexposed to make the secondary maxima show up better.

Huygens' Principle

Our analysis of the superposition of waves from distinct sources, such as two loudspeakers or the two slits in a double-slit experiment, has tacitly assumed that the sources are *point sources,* with no measurable extent. To understand diffraction, we need to think about the propagation of an *extended* wave front. This is a problem that was first considered by the Dutch scientist Christiaan Huygens, a contemporary of Newton who argued that light is a wave.

Huygens lived before a mathematical theory of waves had been developed, so he developed a geometrical model of wave propagation. His idea, which we now call **Huygens' principle,** has two steps:

1. Each point on a wave front is the source of a spherical *wavelet* that spreads out at the wave speed.
2. At a later time, the shape of the wave front is the line tangent to all the wavelets.

Figure 22.11 illustrates Huygens' principle for a plane wave and a spherical wave. As you can see, the line tangent to the wavelets of a plane wave is a plane that has propagated to the right. The line tangent to the wavelets of a spherical wave is a larger sphere. We've shown only a small number of wavelets, which leaves the new surface looking a bit bumpy, but you can easily imagine that the new surface would be perfectly smooth if we drew a wavelet for *every* point on the wave front.

Huygens' principle is a visual device, not a theory of waves. Nonetheless, the full mathematical theory of waves, as it developed in the 19th century, justifies Huygens' basic idea, although it is beyond the scope of this textbook to prove it.

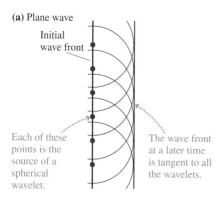

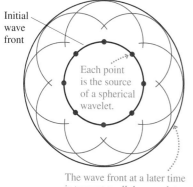

FIGURE 22.11 Huygens' principle applied to the propagation of plane waves and spherical waves.

Analyzing Single-Slit Diffraction

Figure 22.12a shows a wave front passing through a narrow slit of width a. According to Huygens' principle, each point on the wave front can be thought of as the source of a spherical wavelet. These wavelets overlap and interfere, producing the diffraction pattern seen on the viewing screen. The full mathematical analysis, using *every* point on the wave front, is a fairly difficult problem in calculus. We'll be satisfied with a geometrical analysis based on just a few wavelets.

Figure 22.12b shows several wavelets that travel straight ahead to the central point on the screen. (The screen is *very* far to the right in this magnified view of the slit.) The paths to the screen are very nearly parallel to each other, thus all the wavelets travel the same distance and arrive at the screen *in phase* with each other. The *constructive interference* between these wavelets produces the central maximum of the diffraction pattern at $\theta = 0$.

The situation is different at points away from the center. Wavelets 1 and 2 in Figure 22.12c start from points that are distance $a/2$ apart. Suppose that Δr_{12}, the extra distance traveled by wavelet 2, happens to be $\lambda/2$. In that case, wavelets 1 and 2 arrive out of phase and interfere destructively. But if Δr_{12} is $\lambda/2$, then the difference Δr_{34} between paths 3 and 4 and the difference Δr_{56} between paths 5 and 6 are also $\lambda/2$. Those pairs of wavelets also interfere destructively. The superposition of all the wavelets produces perfect destructive interference.

(a) Greatly magnified view of slit

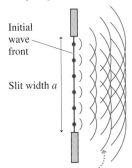

The wavelets from each point on the initial wave front overlap and interfere, creating a diffraction pattern on the screen.

(b)

The wavelets going straight forward all travel the same distance to the screen. Thus they arrive in phase and interfere constructively to produce the central maximum.

(c)

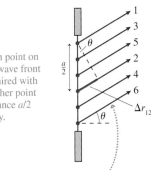

Each point on the wave front is paired with another point distance $a/2$ away.

These wavelets all meet on the screen at angle θ. Wavelet 2 travels distance $\Delta r_{12} = (a/2)\sin\theta$ farther than wavelet 1.

FIGURE 22.12 Each point on the wave front is a source of spherical wavelets. The superposition of these wavelets produces the diffraction pattern on the screen.

Figure 22.12c happens to show six wavelets, but our conclusion is valid for any number of wavelets. The key idea is that **every point on the wave front can be paired with another point that is distance $a/2$ away.** If the path-length difference is $\lambda/2$, the wavelets that originate at these two points will arrive at the screen out of phase and interfere destructively. When we sum the displacements of all N wavelets, they will—pair by pair—add to zero. The viewing screen at this position will be dark. This is the main idea of the analysis, one worth thinking about carefully.

You can see from Figure 22.12c that $\Delta r_{12} = (a/2)\sin\theta$. This path-length difference will be $\lambda/2$, the condition for destructive interference, if

$$\Delta r_{12} = \frac{a}{2}\sin\theta_1 = \frac{\lambda}{2} \tag{22.18}$$

or, equivalently, if $a\sin\theta_1 = \lambda$.

NOTE ▶ Equation 22.18 cannot be satisfied if the slit width a is less than the wavelength λ. If a wave passes through an opening smaller than the wavelength, the central maximum of the diffraction pattern expands to where it *completely* fills the space behind the opening. There are no minima or dark spots at any angle. This situation is uncommon for light waves, because λ is so small, but quite common in the diffraction of sound and water waves. ◀

We can extend this idea to find other angles of perfect destructive interference. Suppose each wavelet is paired with another wavelet from a point $a/4$ away. If Δr between these wavelets is $\lambda/2$, then all N wavelets will again cancel in pairs to give complete destructive interference. The angle θ_2 at which this occurs is found by replacing $a/2$ in Equation 22.18 with $a/4$, leading to the condition $a\sin\theta_2 = 2\lambda$. This process can be continued, and we find that the general condition for complete destructive interference is

$$a\sin\theta_p = p\lambda \qquad p = 1, 2, 3, \ldots \tag{22.19}$$

When $\theta_p \ll 1$ rad, which is almost always true for light waves, we can use the small-angle approximation to write

$$\theta_p = p\frac{\lambda}{a} \qquad p = 1, 2, 3, \ldots \qquad \text{(angles of dark fringes)} \tag{22.20}$$

(a)

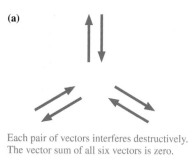

Each pair of vectors interferes destructively. The vector sum of all six vectors is zero.

(b)

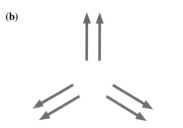

Each pair of vectors interferes constructively. Even so, the vector sum of all six vectors is zero.

FIGURE 22.13 Destructive interference by pairs leads to net destructive interference, but constructive interference by pairs does *not* necessarily lead to net constructive interference.

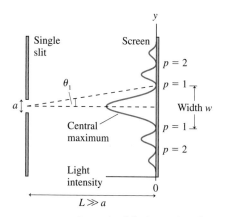
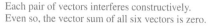

FIGURE 22.14 A graph of the intensity of a single-slit diffraction pattern.

Equation 22.20 gives the angles *in radians* to the dark minima in the diffraction pattern of Figure 22.10. Notice that $p = 0$ is explicitly *excluded*. $p = 0$ corresponds to the straight-ahead position at $\theta = 0$, but you saw in Figures 22.10 and 22.12b that $\theta = 0$ is the central *maximum,* not a minimum.

NOTE ▶ It is perhaps surprising that Equations 22.19 and 22.20 are *mathematically* the same as the condition for the *m*th *maximum* of the double-slit interference pattern. But the physical meaning here is quite different. Equation 22.20 locates the *minima* (dark fringes) of the single-slit diffraction pattern. ◀

You might think that we could use this method of pairing wavelets from different points on the wave front to find the maxima in the diffraction pattern. Why not take two points on the wave front that are distance $a/2$ apart, find the angle at which their wavelets are in phase and interfere constructively, then sum over all points on the wave front? There is a subtle but important distinction. Figure 22.13 shows six vector arrows. The arrows in Figure 22.13a are arranged in pairs such that the two members of each pair cancel. The sum of all six vectors is clearly the zero vector $\vec{0}$, representing destructive interference. This is the procedure we used in Figure 22.12c to arrive at Equation 22.18.

The arrows in Figure 22.13b are arranged in pairs such that the two members of each pair point in the same direction—constructive interference! Nonetheless, the sum of all six vectors is still $\vec{0}$. To have N waves interfere constructively requires more than simply having constructive interference between pairs. Each pair must also be in phase with every other pair, a condition not satisfied in Figure 22.13b. Stated another way, destructive interference by pairs leads to net destructive interference, but constructive interference by pairs does *not* necessarily lead to net constructive interference. It turns out that there is no simple formula to locate the maxima of a single-slit diffraction pattern.

It is possible, although beyond the scope of this textbook, to calculate the entire light intensity pattern. The results of such a calculation are shown graphically in Figure 22.14. You can see the bright central maximum at $\theta = 0$, the weaker secondary maxima, and the dark points of destructive interference at the angles given by Equation 22.20. Compare this graph to the photograph of Figure 22.10 and make sure you see the agreement between the two.

EXAMPLE 22.4 Diffraction of a laser through a slit

Light from a helium-neon laser ($\lambda = 633$ nm) passes through a narrow slit and is seen on a screen 2.0 m behind the slit. The first minimum in the diffraction pattern is 1.2 cm from the central maximum. How wide is the slit?

MODEL A narrow slit produces a single-slit diffraction pattern. A displacement of only 1.2 cm in a distance of 200 cm means that angle θ_1 is certainly a small angle.

VISUALIZE The intensity pattern will look like Figure 22.14.

SOLVE We can use the small-angle approximation to find that the angle to the first minimum is

$$\theta_1 = \frac{1.2 \text{ cm}}{200 \text{ cm}} = 0.00600 \text{ rad} = 0.344°$$

The first minimum is at angle $\theta_1 = \lambda/a$, from which we find that the slit width is

$$a = \frac{\lambda}{\theta_1} = \frac{633 \times 10^{-9} \text{ m}}{6.00 \times 10^{-3} \text{ rad}} = 1.06 \times 10^{-4} \text{ m} = 0.106 \text{ mm}$$

ASSESS This is typical of the slit widths used to observe single-slit diffraction. You can see that the small-angle approximation is well satisfied.

The Width of a Single-Slit Diffraction Pattern

We'll find it useful, as we did for the double slit, to measure positions on the screen rather than angles. The position of the pth dark fringe, at angle θ_p, is $y_p = L\tan\theta_p$, where L is the distance from the slit to the viewing screen. Using Equation 22.20 for θ_p and the small-angle approximation $\tan\theta_p \approx \theta_p$, we find that the dark fringes in the single-slit diffraction pattern are located at

$$y_p = \frac{p\lambda L}{a} \qquad p = 1, 2, 3, \ldots \qquad \text{(positions of dark fringes)} \qquad (22.21)$$

$p = 0$ is explicitly excluded because the midpoint on the viewing screen is the central maximum, not a dark fringe.

A diffraction pattern is dominated by the central maximum, which is much brighter than the secondary maxima. The width w of the central maximum, shown in Figure 22.14, is defined as the distance between the two $p = 1$ minima on either side of the central maximum. Because the pattern is symmetrical, the width is simply $w = 2y_1$. This is

$$w = \frac{2\lambda L}{a} \qquad (22.22)$$

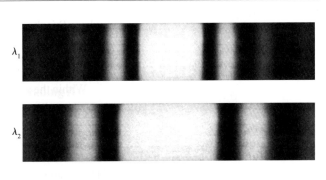

The central maximum of this single-slit diffraction pattern appears white because it is overexposed. The width of the central maximum is clear.

The width of the central maximum is *twice* the spacing $\lambda L/a$ between the dark fringes on either side. The farther back you move the screen (increasing L), the wider the pattern of light on it becomes. In other words, the light waves are *spreading out* behind the slit, and they fill a wider and wider region as they travel farther.

An important implication of Equation 22.22, one contrary to common sense, is that a narrower slit (smaller a) causes a *wider* diffraction pattern. **The smaller the opening you squeeze a wave through, the *more* it spreads out on the other side.**

EXAMPLE 22.5 Determining the wavelength
Light passes through a 0.12-mm-wide slit and forms a diffraction pattern on a screen 1.00 m behind the slit. The width of the central maximum is 0.85 cm. What is the wavelength of the light?

SOLVE From Equation 22.22, the wavelength is

$$\lambda = \frac{aw}{2L} = \frac{(1.2 \times 10^{-4}\,\text{m})(0.0085\,\text{m})}{2(1.00\,\text{m})}$$

$$= 5.10 \times 10^{-7}\,\text{m} = 510\,\text{nm}$$

STOP TO THINK 22.4   The figure shows two single-slit diffraction patterns. The distance between the slit and the viewing screen is the same in both cases. Which of the following could be true?

a. The slits are the same for both; $\lambda_1 > \lambda_2$.
b. The slits are the same for both; $\lambda_2 > \lambda_1$.
c. The wavelengths are the same for both; $a_1 > a_2$.
d. The wavelengths are the same for both; $a_2 > a_1$.
e. The slits and the wavelengths are the same
   for both; $p_1 > p_2$.
f. The slits and the wavelengths are the same
   for both; $p_2 > p_1$.

The physical distance has not changed, but the number of wavelengths along the lower path has. Filling the cell has increased the lower path by

$$\Delta m = m_2 - m_1 = (n - 1)\frac{2d}{\lambda_{vac}} \qquad (22.36)$$

wavelengths. Each increase of one wavelength causes one bright-dark-bright fringe shift at the output, so the index of refraction can be determined by counting fringe shifts as the cell is filled.

EXAMPLE 22.9 Measuring the index of refraction

A Michelson interferometer uses a helium-neon laser with wavelength $\lambda_{vac} = 633.0$ nm. As a 4.00-cm-thick cell is slowly filled with a gas, 43 bright-dark-bright fringe shifts are seen and counted. What is the index of refraction of the gas at this wavelength?

MODEL The gas increases the number of wavelengths in one arm of the interferometer. Each additional wavelength causes one bright-dark-bright fringe shift.

SOLVE We can rearrange Equation 22.36 to find that the index of refraction is

$$n = 1 + \frac{\lambda_{vac}\Delta m}{2d} = 1 + \frac{(6.330 \times 10^{-7}\text{ m})(43)}{2(0.0400\text{ m})} = 1.00034$$

ASSESS This may seem like a six-significant-figure result, but it's really only two. What we're measuring is not n but $n - 1$. We know the fringe count to two significant figures, and that has allowed us to compute $n - 1 = \lambda_{vac}\Delta m/2d = 3.4 \times 10^{-4}$.

Holography

No discussion of wave optics would be complete without mentioning holography, which has both scientific and artistic applications. The basic idea is a simple extension of interferometry.

Figure 22.23a shows how a **hologram** is made. A beam splitter divides a laser beam into two waves. One wave illuminates the object of interest. The light scattered by this object is a very complex wave, but it is the wave you would see if you looked at the object from the position of the film. The other wave, called the *reference beam,* is reflected directly toward the film. The scattered light and the reference beam meet at the film and interfere. The film records their interference pattern.

(a) Recording a hologram

The interference between the scattered light and the reference beam is recorded on the film.

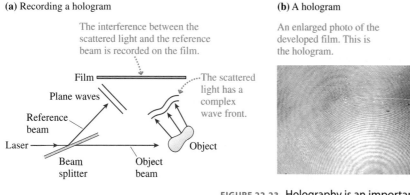

Film · Plane waves · Reference beam · Laser · Beam splitter · Object beam · Object · The scattered light has a complex wave front.

(b) A hologram

An enlarged photo of the developed film. This is the hologram.

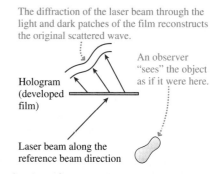

(c) Playing a hologram

The diffraction of the laser beam through the light and dark patches of the film reconstructs the original scattered wave.

Hologram (developed film) · An observer "sees" the object as if it were here. · Laser beam along the reference beam direction

FIGURE 22.23 Holography is an important application of wave optics.

The interference patterns we've looked at in this chapter have been simple geometrical patterns of stripes and circles because the light waves have been well-behaved plane waves and spherical waves. The light wave scattered by the object in Figure 22.23a is exceedingly complex. As a result, the interference pattern recorded on the film—the hologram—is a seemingly random pattern of whorls and blotches. Figure 22.23b is an enlarged photograph of a portion of a hologram. It's certainly not obvious that information is stored in this pattern, but it is.

The hologram is "played" by sending just the reference beam through it, as seen in Figure 22.23c. The reference beam diffracts through the transparent parts of the hologram, just as it would through the slits of a diffraction grating. Amazingly, the diffracted wave is *exactly the same* as the light wave that had been scattered by the object! In other words, the diffracted reference beam *reconstructs* the original scattered wave. As you look at this diffracted wave, from the far side of the hologram, you "see" the object exactly as if it were there. The view is three dimensional because, by moving your head with respect to the hologram, you can see different portions of the wave front.

SUMMARY

The goal of Chapter 22 has been to understand and apply the wave model of light.

GENERAL PRINCIPLES

Huygens' principle says that each point on a wave front is the source on a spherical wavelet. The wave front at a later time is tangent to all the wavelets.

Diffraction is the spreading of a wave after it passes through an opening.

Constructive and destructive interference are due to the overlap of two or more waves as they spread behind openings.

IMPORTANT CONCEPTS

The wave model of light considers light to be a wave propagating through space. Diffraction and interference are important.

The ray model of light considers light to travel in straight lines like little particles. Diffraction and interference are not important.

Diffraction is important when the width of the diffraction pattern of an aperture equals or exceeds the size of the aperture. For a circular aperture, the crossover between the ray and wave models occurs for an opening of diameter $D_c = \sqrt{2.44\lambda L}$.

In practice, $D_c \approx 1$ mm. Thus

- Use the wave model when light passes through openings <1 mm in size. Diffraction effects are usually important.
- Use the ray model when light passes through openings >1 mm in size. Diffraction is usually not important.

APPLICATIONS

Single slit of width a.
A bright **central maximum** of width

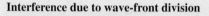

$$w = \frac{2\lambda L}{a}$$

is flanked by weaker **secondary maxima.** Dark fringes are located at angles such that

$$a\sin\theta_p = p\lambda \qquad p = 1, 2, 3, \ldots$$

If $\lambda/p \ll 1$, then from the small-angle approximation

$$\theta_p = \frac{p\lambda}{a} \qquad y_p = \frac{p\lambda L}{a}$$

Interference due to wave-front division

Waves overlap as they spread out behind slits. Constructive interference occurs along antinodal lines. Bright fringes are seen where the antinodal lines intersect the viewing screen.

Double slit with separation d.
Equally spaced bright fringes are located at

$$\theta_m = \frac{m\lambda}{d} \qquad y_m = \frac{m\lambda L}{d} \qquad m = 0, 1, 2, \ldots$$

The **fringe spacing** is $\Delta y = \frac{\lambda L}{d}$

Diffraction grating with slit spacing d.
Very bright and narrow fringes are located at angles and positions

$$d\sin\theta_m = m\lambda \qquad y_m = L\tan\theta_m$$

Circular aperture of diameter D.
A bright central maximum of diameter

$$w = \frac{2.44\lambda L}{D}$$

is surrounded by circular secondary maxima. The first dark fringe is located at

$$\theta_1 = \frac{1.22\lambda}{D} \qquad y_1 = \frac{1.22\lambda L}{D}$$

For an aperture of any shape, a smaller opening causes a more rapid spreading of the wave behind the opening.

Interference due to amplitude division

An interferometer divides a wave, lets the two waves travel different paths, then recombines them. Interference is constructive if one wave travels an integer number of wavelengths more or less than the other wave. The difference can be due to an actual path-length difference or to a different index of refraction.

Michelson interferometer

The number of bright-dark-bright fringe shifts as mirror M_2 moves distance ΔL_2 is

$$\Delta m = \frac{\Delta L_2}{\lambda/2}$$

TERMS AND NOTATION

optics	double slit	order, m	circular aperture
diffraction	interference fringes	spectroscopy	interferometer
models of light	central maximum	single-slit diffraction	beam splitter
wave optics	fringe spacing, Δy	secondary maxima	hologram
ray optics	diffraction grating	Huygens' principle	

EXERCISES AND PROBLEMS

Exercises

Section 22.2 The Interference of Light

1. Figure Ex22.1 shows light waves passing through two closely spaced slits. The graph shows the light intensity on a screen behind the slits. Reproduce the graph axes on your page, including the zero and the tick marks that locate the fringes of the double-slit interference pattern.
   a. Draw a graph to show how the light intensity pattern will appear if the right slit is blocked, allowing light to go only through the left slit.
   b. Explain why you drew the graph this way.

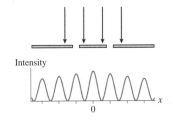

FIGURE EX22.1

2. Light of wavelength 500 nm illuminates a double slit, and the interference pattern is observed on a screen. At the position of the $m = 2$ bright fringe, how much farther is it to the more distant slit than to the nearer slit?
3. Two narrow slits 50 μm apart are illuminated with light of wavelength 500 nm. What is the angle of the $m = 2$ bright fringe in radians? In degrees?
4. Light from a sodium lamp ($\lambda = 589$ nm) illuminates two narrow slits. The fringe spacing on a screen 150 cm behind the slits is 4.0 mm. What is the spacing (in mm) between the two slits?
5. A double-slit interference pattern is created by two narrow slits spaced 0.20 mm apart. The distance between the first and the fifth minimum on a screen 60 cm behind the slits is 6.0 mm. What is the wavelength of the light used in this experiment?
6. A double-slit experiment is performed with light of wavelength 600 nm. The bright interference fringes are spaced 1.8 mm apart on the viewing screen. What will the fringe spacing be if the light is changed to a wavelength of 400 nm?
7. Light from a helium-neon laser ($\lambda = 633$ nm) is used to illuminate two narrow slits. The interference pattern is observed on a screen 3.0 m behind the slits. Twelve bright fringes are seen, spanning a distance of 52 mm. What is the spacing (in mm) between the slits?

Section 22.3 The Diffraction Grating

8. A 1.0-cm-wide diffraction grating has 1000 slits. It is illuminated by light of wavelength 550 nm. What are the angles of the first two diffraction orders?
9. Light of wavelength 600 nm illuminates a diffraction grating. The second-order maximum is at angle 39.5°. How many lines per millimeter does this grating have?
10. A helium-neon laser ($\lambda = 633$ nm) illuminates a diffraction grating. The distance between the two $m = 1$ bright fringes is 32 cm on a screen 2.0 m behind the grating. What is the spacing between slits of the grating?
11. A diffraction grating with 600 lines/mm is illuminated with light of wavelength 500 nm. A very wide viewing screen is 2.0 m behind the grating.
    a. What is the distance between the two $m = 1$ bright fringes?
    b. How many bright fringes can be seen on the screen?
12. A 500 line/mm diffraction grating is illuminated by light of wavelength 510 nm. How many diffraction orders are seen, and what is the angle of each?
13. The two most prominent wavelengths in the light emitted by a hydrogen discharge lamp are 656 nm (red) and 486 nm (blue). Light from a hydrogen lamp illuminates a diffraction grating with 500 lines/mm, and the light is observed on a screen 1.5 m behind the grating. What is the distance between the first-order red and blue fringes?

Section 22.4 Single-Slit Diffraction

14. A helium-neon laser ($\lambda = 633$ nm) illuminates a single slit and is observed on a screen 1.5 m behind the slit. The distance between the first and second minima in the diffraction pattern is 4.75 mm. What is the width (in mm) of the slit?
15. A 0.50-mm-wide slit is illuminated by light of wavelength 500 nm. What is the width of the central maximum on a screen 2.0 m behind the slit?
16. In a single-slit experiment, the slit width is 200 times the wavelength of the light. What is the width of the central maximum on a screen 2.0 m behind the slit?
17. The second minimum in the diffraction pattern of a 0.10-mm-wide slit occurs at 0.70°. What is the wavelength of the light?
18. You need to use your cell phone, which broadcasts an 800 MHz signal, but you're behind two massive, radio-wave-absorbing buildings that have only a 15 m space between them. What is the angular width, in degrees, of the electromagnetic wave after it emerges from between the buildings?

19. The opening to a cave is a tall, 30-cm-wide crack. A bat that is preparing to leave the cave emits a 30 kHz ultrasonic chirp. How wide is the "sound beam" 100 m outside the cave opening? Use $v_{\text{sound}} = 340$ m/s.

Section 22.5 Circular-Aperture Diffraction

20. A 0.50-mm-diameter hole is illuminated by light of wavelength 500 nm. What is the width of the central maximum on a screen 2.0 m behind the slit?
21. Light from a helium-neon laser ($\lambda = 633$ nm) passes through a circular aperture and is observed on a screen 4.0 m behind the aperture. The width of the central maximum is 2.5 cm. What is the diameter (in mm) of the hole?
22. You want to photograph a circular diffraction pattern whose central maximum has a diameter of 1.0 cm. You have a helium-neon laser ($\lambda = 633$ nm) and a 0.12-mm-diameter pinhole. How far behind the pinhole should you place the viewing screen?
23. Infrared light of wavelength 2.5 μm illuminates a 0.20-mm-diameter hole. What is the angle of the first dark fringe in radians? In degrees?

Section 22.6 Interferometers

24. Moving mirror M_2 of a Michelson interferometer a distance of 100 μm causes 500 bright-dark-bright fringe shifts. What is the wavelength of the light?
25. A Michelson interferometer uses red light with a wavelength of 656.45 nm from a hydrogen discharge lamp. How many bright-dark-bright fringe shifts are observed if mirror M_2 is moved exactly 1 cm?
26. A Michelson interferometer uses light whose wavelength is known to be 602.446 nm. Mirror M_2 is slowly moved while exactly 33,198 bright-dark-bright fringe shifts are observed. What distance has M_2 moved? Be sure to give your answer to an appropriate number of significant figures.
27. A Michelson interferometer operating at a 600 nm wavelength has a 2.00-cm-long glass cell in one arm. To begin, the air is pumped out of the cell and mirror M_2 is adjusted to produce a bright spot at the center of the interference pattern. Then a valve is opened and air is slowly admitted into the cell. The index of refraction of air at 1.0 atm pressure is 1.00028. How many bright-dark-bright fringe shifts are observed as the cell fills with air?

Problems

28. Figure P22.28 shows the light intensity on a screen 2.5 m behind an aperture. The aperture is illuminated with light of wavelength 600 nm.
    a. Is the aperture a single slit or a double slit? Explain.
    b. If the aperture is a single slit, what is its width? If it is a double slit, what is the spacing between the slits?

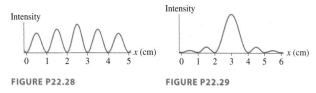

FIGURE P22.28   **FIGURE P22.29**

29. Figure P22.29 shows the light intensity on a screen 2.5 m behind an aperture. The aperture is illuminated with light of wavelength 600 nm.
    a. Is the aperture a single slit or a double slit? Explain.
    b. If the aperture is a single slit, what is its width? If it is a double slit, what is the spacing between the slits?
30. A double slit is illuminated simultaneously with orange light of wavelength 600 nm and light of an unknown wavelength. The $m = 4$ bright fringe of the unknown wavelength overlaps the $m = 3$ bright orange fringe. What is the unknown wavelength?
31. In a double-slit experiment, the slit separation is 200 times the wavelength of the light. What is the angular separation between two adjacent bright fringes?
32. What is the intensity of a double-slit interference pattern at a point on the screen halfway between the central maximum and the first minimum? Give your answer as a multiple of I_1, the intensity due to each individual wave.
33. A double-slit interference pattern has a fringe spacing of 4.0 mm. How far from the central maximum is the first position at which the intensity is equal to I_1?
34. Figure P22.34 shows the light intensity on a screen behind a double slit. The slit spacing is 0.20 mm and the wavelength of the light is 600 nm. What is the distance from the slits to the screen?

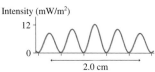

FIGURE P22.34

35. Figure P22.34 shows the light intensity on a screen behind a double slit. The slit spacing is 0.20 mm and the screen is 2.0 m behind the slits. What is the wavelength of the light?
36. Figure P22.34 shows the light intensity on a screen behind a double slit. Suppose one slit is covered. What will be the light intensity at the center of the screen due to the remaining slit?
37. A diffraction grating having 500 lines/mm diffracts visible light at 30°. What is the light's wavelength?
38. Light emitted by Element X passes through a diffraction grating having 1200 lines/mm. The diffraction pattern is observed on a screen 75.0 cm behind the grating. Bright fringes are *seen* on the screen at distances of 56.2 cm, 65.9 cm, and 93.5 cm from the central maximum. No other fringes are seen.
    a. What is the value of m for each of these diffracted wavelengths? Explain why only one value is possible.
    b. What are the wavelengths of light emitted by Element X?
39. Helium atoms emit light at several wavelengths. Light from a helium lamp illuminates a diffraction grating and is observed on a screen 50.0 cm behind the grating. The emission at wavelength 501.5 nm creates a first-order bright fringe 21.90 cm from the central maximum. What is the wavelength of the bright fringe that is 31.60 cm from the central maximum?
40. A diffraction grating produces a first-order maximum at an angle of 20.0°. What is the angle of the second-order maximum?
41. A diffraction grating is illuminated simultaneously with red light of wavelength 660 nm and light of an unknown wavelength. The fifth-order maximum of the unknown wavelength exactly overlaps the third-order maximum of the red light. What is the unknown wavelength?

42. White light (400–700 nm) is incident on a 600 line/mm diffraction grating. What is the width of the first-order rainbow on a screen 2.0 m behind the grating?

43. White light (400–700 nm) is incident on a 300 line/mm diffraction grating. What wavelengths in the third-order rainbow overlap wavelengths in the fourth-order rainbow?

44. For your science fair project you need to design a diffraction grating that will disperse the visible spectrum (400–700 nm) over 30.0° in first order.
    a. How many lines per millimeter does your grating need?
    b. What is the first-order diffraction angle of light from a sodium lamp ($\lambda = 589$ nm)?

45. Figure P22.45 shows the interference pattern on a screen 1.0 m behind an 800 line/mm diffraction grating. What is the wavelength of the light?

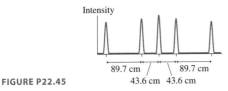

Intensity

89.7 cm   89.7 cm
43.6 cm 43.6 cm

FIGURE P22.45

46. Figure P22.45 shows the interference pattern on a screen 1.0 m behind a diffraction grating. The wavelength of the light is 600 nm. How many lines per millimeter does the grating have?

47. Light from a sodium lamp ($\lambda = 589$ nm) illuminates a narrow slit and is observed on a screen 75 cm behind the slit. The distance between the first and third dark fringes is 7.5 mm. What is the width (in mm) of the slit?

48. What is the width of a slit for which the first minimum is at 45° when the slit is illuminated by a helium-neon laser ($\lambda = 633$ nm)?

49. For what slit-width-to-wavelength ratio does the first minimum of a single-slit diffraction pattern appear at (a) 30°, (b) 60°, and (c) 90°?

50. Light from a helium-neon laser ($\lambda = 633$ nm) is incident on a single slit. What is the largest slit width for which there are no minima in the diffraction pattern?

51. Figure P22.51 shows the light intensity on a screen behind a single slit. The wavelength of the light is 500 nm and the screen is 1.0 m behind the slit. What is the width (in mm) of the slit?

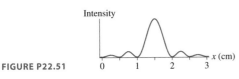

Intensity

x (cm)
0   1   2   3

FIGURE P22.51

52. Figure P22.51 shows the light intensity on a screen behind a single slit. The wavelength of the light is 600 nm and the slit width is 0.15 mm. What is the distance from the slit to the screen?

53. Figure P22.51 shows the light intensity on a screen behind a circular aperture. The wavelength of the light is 500 nm and the screen is 1.0 m behind the slit. What is the diameter (in mm) of the aperture?

54. Light from a helium-neon laser ($\lambda = 633$ nm) illuminates a circular aperture. It is noted that the diameter of the central maximum on a screen 50 cm behind the aperture matches the diameter of the geometric image. What is the aperture's diameter?

55. One day, after pulling down your window shade, you notice that sunlight is passing through a pinhole in the shade and making a small patch of light on the far wall. Having recently studied optics in your physics class, you're not too surprised to see that the patch of light seems to be a circular diffraction pattern. It appears that the central maximum is about 1 cm across, and you estimate that the distance from the window shade to the wall is about 3 m. Estimate (a) the average wavelength of sunlight and (b) the diameter of the pinhole.

56. A radar for tracking aircraft broadcasts a 12 GHz microwave beam from a 2.0-m-diameter circular radar antenna. From a wave perspective, the antenna is a circular aperture through which the microwaves diffract.
    a. What is the diameter of the radar beam at a distance of 30 km?
    b. If the antenna emits 100 kW of power, what is the average microwave intensity at 30 km?

57. A helium-neon laser ($\lambda = 633$ nm) is built with a glass tube of inside diameter 1.0 mm. One mirror is partially transmitting to allow the laser beam out. An electrical discharge in the tube causes it to glow like a neon light. From an optical perspective, the laser beam is a light wave that diffracts out through a 1.0-mm-diameter circular opening.
    a. Can a laser beam be *perfectly* parallel, with no spreading? Why or why not?
    b. The angle θ_1 to the first minimum is called the *divergence angle* of a laser beam. What is the divergence angle of this laser beam?
    c. What is the diameter (in mm) of the laser beam after it travels 3.0 m?

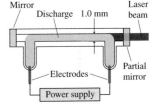

Mirror   Laser
Discharge 1.0 mm   beam

Partial
Electrodes   mirror

Power supply

FIGURE P22.57

    d. What is the diameter of the laser beam after it travels 1.0 km?

58. Scientists use *laser range-finding* to measure the distance to the moon with great accuracy. A brief laser pulse is fired at the moon, then the time interval is measured until the "echo" is seen by a telescope. A laser beam spreads out as it travels because it diffracts through a circular exit as it leaves the laser. In order for the reflected light to be bright enough to detect, the laser spot on the moon must be no more than 1 km in diameter. Staying within this diameter is accomplished by using a special large-diameter laser. If $\lambda = 532$ nm, what is the minimum diameter of the circular opening from which the laser beam emerges? The earth-moon distance is 384,000 km.

59. Light of wavelength 600 nm passes though two slits separated by 0.20 mm and is observed on a screen 1.0 m behind the slits. The location of the central maximum is marked on the screen and labeled $y = 0$.
    a. At what distance, on either side of $y = 0$, are the $m = 1$ bright fringes?
    b. A very thin piece of glass is then placed in one slit. Because light travels slower in glass than in air, the wave passing through the glass is delayed by 5.0×10^{-16} s in comparison to the wave going through the other slit. What fraction of the period of the light wave is this delay?
    c. With the glass in place, what is the phase difference $\Delta\phi_0$ between the two waves as they leave the slits?
    d. The glass causes the interference fringe pattern on the screen to shift sideways. Which way does the central maximum move (toward or away from the slit with the glass) and by how far?

60. A 600 line/mm diffraction grating is in an empty aquarium tank. The index of refraction of the glass walls is $n_{glass} = 1.50$. A helium-neon laser ($\lambda = 633$ nm) is outside the aquarium. The laser beam passes through the glass wall and illuminates the diffraction grating.
    a. What is the first-order diffraction angle of the laser beam?
    b. What is the first-order diffraction angle of the laser beam after the aquarium is filled with water ($n_{water} = 1.33$)?

61. You've set up a Michelson interferometer with a helium-neon laser ($\lambda = 632.8$ nm). After adjusting mirror M_2 to produce a bright spot at the center of the pattern, you carefully move M_2 away from the beam splitter while counting 1200 new bright spots at the center. Then you put the laser away. Later another student wants to restore the interferometer to its starting condition, but he mistakenly sets up a hydrogen discharge lamp and uses the 656.5 nm emission from hydrogen atoms. He then counts 1200 new bright spots while slowly moving M_2 back toward the beam splitter. What is the net displacement of M_2 when he is done? Is M_2 now closer to or farther from the beam splitter?

62. A Michelson interferometer uses light from a sodium lamp. Sodium atoms emit light having wavelengths 589.0 nm and 589.6 nm. The interferometer is initially set up with both arms of equal length ($L_1 = L_2$), producing a bright spot at the center of the interference pattern. How far must mirror M_2 be moved so that one wavelength has produced one more new maxima than the other wavelength?

63. A light wave has wavelength 500 nm in vacuum.
    a. What is the wavelength of this light as it travels through water ($n_{water} = 1.33$)?
    b. Suppose that a 1.0-mm-thick layer of water is inserted into one arm of a Michelson interferometer. How many "extra" wavelengths does the light now travel in this arm?
    c. By how many fringes will this water layer shift the interference pattern?

64. A 0.10-mm-thick piece of glass is inserted into one arm of a Michelson interferometer that is using light of wavelength 500 nm. This causes the fringe pattern to shift by 200 fringes. What is the index of refraction of this piece of glass?

65. Transparent materials known as *electro-optic crystals* change their index of refraction when a voltage is applied to them. They are used to control the output of interferometers in optical computers and optical switching devices. To illustrate the basic idea, suppose one arm of a Michelson interferometer contains a 0.100-mm-thick electro-optic crystal with an initial index of refraction $n = 1.550$. The interferometer operates with light of wavelength 0.981 μm from a semiconductor laser. Mirror M_2 is first adjusted to make the output a bright fringe, which we can interpret as a 1 in binary arithmetic. What is the first index of refraction larger than 1.550 for which output goes dark—a binary 0? A voltage that can change the index of refraction by this amount can be used to switch binary signals from 1's to 0's or from 0's to 1's.

66. To illustrate one of the ideas of holography in a simple way, consider a diffraction grating with slit spacing d. The small-angle approximation is usually not valid for diffraction gratings, because d is only slightly larger than λ, but assume that the λ/d ratio of this grating is small enough to make the small-angle approximation valid.
    a. Use the small-angle approximation to find an expression for the fringe spacing on a screen at distance L behind the grating.
    b. Rather than a screen, suppose you place a piece of film at distance L behind the grating. The bright fringes will expose the film, but the dark spaces in between will leave the film unexposed. After being developed, the film will be a series of alternating light and dark stripes. What if you were to now "play" the film by using it as a diffraction grating? In other words, what happens if you shine the same laser through the film and look at the film's diffraction pattern on a screen at the same distance L? Demonstrate that the film's diffraction pattern is a reproduction of the original diffraction grating.

Challenge Problems

67. A double-slit experiment is set up using a helium-neon laser ($\lambda = 633$ nm). Then a very thin piece of glass ($n = 1.50$) is placed over one of the slits. Afterward, the central point on the screen is occupied by what had been the $m = 10$ dark fringe. How thick is the glass?

68. Light of wavelength 600 nm passes through a double slit and is viewed on a screen 2.0 m behind the slits. Each slit is 0.040 mm wide and they are separated by 0.200 mm. How many bright fringes are seen on the screen?
    Hint: What determines the width of the region on the screen where fringes are observed?

69. Light consisting of two nearly equal wavelengths $\lambda + \Delta\lambda$ and λ, where $\Delta\lambda \ll \lambda$, is incident on a diffraction grating. The slit separation of the grating is d.
    a. Show that the angular separation of these two wavelengths in the mth order is
    $$\Delta\theta = \frac{\Delta\lambda}{\sqrt{(d/m)^2 - \lambda^2}}$$
    b. Sodium atoms emit light at 589.0 nm and 589.6 nm. What are the first-order and second-order angular separations (in degrees) of these two wavelengths for a 600 line/mm grating?

70. Figure CP22.70 shows two nearly overlapped intensity peaks of the sort you might produce with a diffraction grating (see Figure 22.8b). As a practical matter, two peaks can just barely be resolved if their spacing Δy equals the width w of each peak, where w is measured at half of the peak's height. Two peaks closer together than w will merge into a single peak. We can use this idea to understand the *resolution* of a diffraction grating.

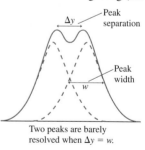

Two peaks are barely resolved when $\Delta y = w$.

FIGURE CP22.70

    a. In the small-angle approximation, the position of the $m = 1$ peak of a diffraction grating falls at the same location as the $m = 1$ fringe of a double slit: $y_1 = \lambda L/d$. Suppose two wavelengths differing by $\Delta\lambda$ pass through a grating at the same time. Find an expression for Δy, the separation of their first-order peaks.
    b. We noted that the widths of the bright fringes are proportional to $1/N$, where N is the number of slits in the grating. Let's hypothesize that the fringe width is $w = y_1/N$. Show

that this is true for the double-slit pattern. We'll then assume it to be true as N increases.

c. Use your results from parts a and b together with the idea that $\Delta y_{min} = w$ to find an expression for $\Delta\lambda_{min}$, the minimum wavelength separation (in first order) for which the diffraction fringes can barely be resolved.

d. Ordinary hydrogen atoms emit red light with a wavelength of 656.45 nm. In deuterium, which is a "heavy" isotope of hydrogen, the wavelength is 656.27 nm. What is the minimum number of slits in a diffraction grating that can barely resolve these two wavelengths in the first-order diffraction pattern?

71. The diffraction grating analysis in this chapter assumed that the incident light is normal to the grating. Figure CP22.71 shows a plane wave approaching a diffraction grating at angle ϕ.

a. Show that the angles θ_m for constructive interference are given by the grating equation

$$d(\sin\theta_m + \sin\phi) = m\lambda$$

where $m = 0, \pm 1, \pm 2, \ldots$. Angles are considered positive if they are above the horizontal line, negative if below it.

b. The two first-order maxima, $m = +1$ and $m = -1$, are no longer symmetrical about the center. Find θ_1 and θ_{-1} for 500 nm light incident on a 600 line/mm grating at $\phi = 30°$.

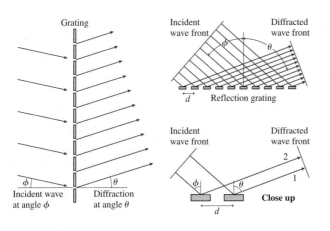

FIGURE CP22.71   **FIGURE CP22.72**

72. Figure CP22.72 shows light of wavelength λ incident at angle ϕ on a *reflection* grating of spacing d. We want to find the angles θ_m at which constructive interference occurs.

a. The figure shows paths 1 and 2 along which two waves travel and interfere. Find an expression for the path-length difference $\Delta r = r_2 - r_1$.

b. Using your result from part a, find an equation (analogous to Equation 22.15) for the angles θ_m at which diffraction occurs when the light is incident at angle ϕ. Notice that m can be a negative integer in your expression, indicating that path 2 is shorter than path 1.

c. Show that the zeroth-order diffraction is simply a "reflection." That is, $\theta_0 = \phi$.

d. Light of wavelength 500 nm is incident at $\phi = 40°$ on a reflection grating having 700 reflection lines/mm. Find all angles θ_m at which light is diffracted. Negative values of θ_m are interpreted as an angle left of the vertical.

e. Draw a picture showing a *single* 500 nm light ray incident at $\phi = 40°$ and showing all the diffracted waves at the correct angles.

73. The pinhole camera of Figure CP22.73 images distant objects by allowing only a narrow bundle of light rays to pass through the hole and strike the film. If light consisted of particles, you could make the image sharper and sharper (at the expense of getting dimmer and dimmer) by making the aperture smaller and smaller. In practice, diffraction of light by the circular aperture limits the maximum sharpness that can be obtained. Consider two distant points of light, such as two distant street-

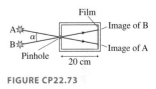

FIGURE CP22.73

lights. Each will produce a circular diffraction pattern on the film. The two images can just barely be resolved if the central maximum of one image falls on the first dark fringe of the other image. (This is called Rayleigh's criterion, and we will explore its implication for optical instruments in the next chapter.)

a. Optimum sharpness of one image occurs when the diameter of the central maximum equals the diameter of the pinhole. What is the optimum hole size for a pinhole camera in which the film is 20 cm behind the hole? Assume $\lambda = 550$ nm, an average value for visible light.

b. For this hole size, what is the angle α (in degrees) between two distant sources that can barely be resolved?

c. What is the distance between two street lights 1 km away that can barely be resolved?

STOP TO THINK ANSWERS

Stop to Think 22.1: b. The antinodal lines seen in Figure 22.3b are diverging.

Stop to Think 22.2: Smaller. Shorter-wavelength light doesn't spread as rapidly as longer-wavelength light. The fringe spacing Δy is directly proportional to the wavelength λ.

Stop to Think 22.3: d. Larger wavelengths have larger diffraction angles. Red light has a larger wavelength than violet light, so red light is diffracted farther from the center.

Stop to Think 22.4: b or c. The width of the central maximum, which is proportional to λ/a, has increased. This could occur either because the wavelength has increased or because the slit width has decreased.

Stop to Think 22.5: d. Moving M_1 in by λ decreases r_1 by 2λ. Moving M_2 out by λ increases r_2 by 2λ. These two actions together change the path length by $\Delta r = 4\lambda$.

23 Ray Optics

A focused laser beam can cut and drill the hardest metals.

Humans have always been fascinated by light. Simple mirrors are found in ancient archeological sites from Egypt to China. Our ancestors had learned by 1500 BCE to start fires by focusing sunlight with a simple lens. From there, it's only a small step to drilling holes with a focused laser beam.

Chapter 22 introduced the three models of light but then emphasized the wave optics of interference and diffraction. This chapter will analyze basic optical systems such as mirrors and lenses in terms of straight-line light trajectories. This is *ray optics,* and it is a subject of immense practical value. The ray model of light will take center stage, but we'll find that we can't entirely avoid the fact that light is a wave. It will turn out, perhaps surprisingly, that the waviness of light is what ultimately sets the performance limits of optical systems. You'll finish this chapter with a much more complete understanding of what light is and how it behaves.

23.1 The Ray Model of Light

A flashlight makes a beam of light through the night's darkness. Sunbeams stream into a darkened room through a small hole in the shade, and laser beams are even more well defined. Our everyday experience that light travels in straight lines is the basis of the *ray model* of light.

The ray model is an oversimplification of reality, but nonetheless is very useful within its range of validity. In particular, the ray model of light is valid as long as any apertures through which the light passes (lenses, mirrors, holes, and the like) are very large compared to the wavelength of light. With such apertures, diffraction and other wave aspects of light are negligible and can be ignored. The analysis of Section 22.5 found that the crossover between wave optics and ray optics occurs for apertures ≈ 1 mm in diameter. Lenses and mirrors are almost always larger than 1 mm, so the ray model of light is an excellent basis for the practical optics of image formation.

To begin, let us define a **light ray** as a line in the direction along which light energy is flowing. A light ray is an abstract idea, not a physical entity or a "thing." Any narrow beam of light, such as the laser beam in Figure 23.1, is actually a bundle of many parallel light rays. You can think of a single light ray as the limiting case of a laser beam whose diameter approaches zero. Laser beams are good approximations of light rays, certainly adequate for demonstrating ray behavior, but any real laser beam is a bundle of many parallel rays.

The following table outlines five basic ideas and assumptions of the ray model of light.

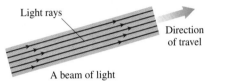

FIGURE 23.1 A laser beam or beam of sunlight is a bundle of parallel light rays.

The ray model of light

(diagram of straight line with arrow)	Light rays travel in straight lines. Light travels through a transparent material in straight lines called light rays. The speed of light is $v = c/n$, where n is the index of refraction of the material.
(diagram of two crossing lines)	Light rays can cross. Light rays do not interact with each other. Two rays can cross without either being affected in any way.
	A light ray travels forever unless it interacts with matter. A light ray continues forever unless it has an interaction with matter that causes the ray to change direction or to be absorbed. Light interacts with matter in four different ways: ■ At an interface between two materials, light can be either *reflected* or *refracted*. ■ Within a material, light can be either *scattered* or *absorbed*. These interactions are discussed later in the chapter.
	An object is a source of light rays. An **object** is a source of light rays. Rays originate from *every* point on the object, and each point sends rays in *all* directions. We make no distinction between self-luminous objects and reflective objects.
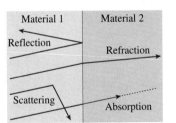 Diverging bundle of rays ... Eye ... d	The eye sees by focusing a diverging bundle of rays. The eye "sees" an object when *diverging* bundles of rays from each point on the object enter the pupil and are focused to an image on the retina. (Imaging is discussed later in the chapter.) From the movements the eye's lens has to make to focus the image, your brain "computes" the distance d at which the rays originated, and you perceive the object as being at that point.

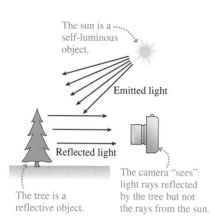

FIGURE 23.2 Self-luminous and reflective objects.

FIGURE 23.3 Point sources and parallel bundles represent idealized objects.

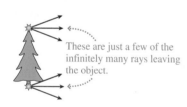

FIGURE 23.4 A ray diagram simplifies the situation by showing only a few rays.

Objects

Figure 23.2 illustrates the idea that objects can be either *self-luminous,* such as the sun, flames, and light bulbs, or *reflective.* Most objects are reflective. A tree, unless it is on fire, is seen or photographed by virtue of reflected sunlight or reflected skylight. People, houses, and this page in the book reflect light from self-luminous sources. In this chapter we are concerned not with how the light originates but with how it behaves after leaving the object.

Light rays from an object are emitted in all directions, but you are not *aware* of light rays unless they enter the pupil of your eye. Consequently, most light rays go completely unnoticed. For example, light rays travel from the sun to the tree in Figure 23.2, but you're not aware of these unless the tree reflects some of them into your eye. Or consider a laser beam traveling through the room. You've probably noticed that it's almost impossible to see a laser beam from the side unless there's dust in the air. The dust scatters a few of the light rays toward your eye, but in the absence of dust you would be completely unaware of a very powerful light beam traveling past you. **Light rays exist everywhere, quite independently of whether you are seeing them.**

Figure 23.3 shows two idealized sets of light rays. The diverging rays from a **point source** are emitted in all directions. It will often be useful to think of each point on an object as a point source of light rays. A **parallel bundle** of rays could be a laser beam. Alternatively it could represent a *distant object,* an object such as a star so far away that the rays arriving at the observer are essentially parallel to each other.

Ray Diagrams

Rays originate from *every* point on an object and travel outward in *all* directions, but a diagram trying to show all these rays would be hopelessly messy and confusing. To simplify the picture, we usually use a **ray diagram** that shows only a few rays. For example, Figure 23.4 is a ray diagram showing only a few rays leaving the top and bottom points of the object and traveling to the right. These rays will be sufficient to show us how the object is imaged by lenses or mirrors.

NOTE ▶ Ray diagrams are the basis for a *pictorial representation* that we'll use throughout this chapter. Be careful not think that a ray diagram shows all of the rays. The rays shown on the diagram are just a subset of the infinitely many rays leaving the object. ◀

Apertures

A popular form of entertainment during ancient Roman times was a visit to a **camera obscura,** Latin for "dark room." As Figure 23.5a shows, a camera obscura was a darkened room with a single, small hole to the outside world. After their eyes became dark adapted, visitors could see a dim but full-color image of

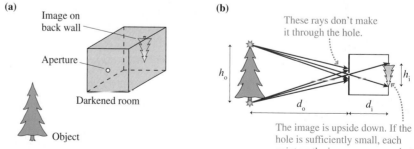

FIGURE 23.5 A camera obscura.

the outside world displayed on the back wall of the room. However, the image was upside down! The *pinhole camera* is a miniature and more modern version of the camera obscura.

A hole through which light passes is called an **aperture.** Figure 23.5b uses the ray model of light passing through a small aperture to explain how the camera obscura works. Each point on an object emits light rays in all directions, but only a very few of these rays pass through the aperture and reach the back wall. As the figure illustrates, the geometry of the rays causes the image to be upside down.

Actually, as you may have realized, each *point* on the object illuminates a small but finite *patch* on the wall. This is because the finite size of the aperture allows several rays from each point on the object to pass through at slightly different angles. As a result, the image is slightly blurred and out of focus. Maximum sharpness is achieved by making the hole smaller and smaller, which makes the image dimmer and dimmer. (Diffraction also becomes an issue if the hole gets too small.) A practical camera obscura or pinhole camera has to accept a small amount of blurring as the tradeoff for having an image bright enough to see.

An interesting aspect of the camera obscura is that the image size is not the same as the object size. We define the **magnification** m as the ratio of image height h_i to object height h_o:

$$m = \frac{h_i}{h_o} \tag{23.1}$$

While you may think of magnification as meaning "larger," our definition allows the image to be either larger than the object ($m > 1$) or smaller ($m < 1$). In the case of the camera obscura, you can see from the similar triangles in Figure 23.5b that

$$m = \frac{h_i}{h_o} = \frac{d_i}{d_o} \tag{23.2}$$

where d_o is the distance to the object and d_i is the depth of the camera obscura. Any realistic camera obscura has $d_i < d_o$ and magnification $m < 1$.

These ideas will all reappear when we consider forming images with lenses. There we will discover how a modern camera, with a lens, improves on the camera obscura.

We can apply the ray model to more complex apertures, such as the L-shaped aperture in Figure 23.6. The pattern of light on the screen is found by tracing all the straight-line paths—the ray trajectories—that start from the point source and pass through the aperture. We will see an enlarged L on the screen, with a sharp boundary between the image and the dark shadow.

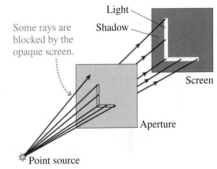

FIGURE 23.6 A point source and an aperture.

STOP TO THINK 23.1   A long, thin light bulb illuminates a vertical aperture. Which pattern of light do you see on a viewing screen behind the aperture?

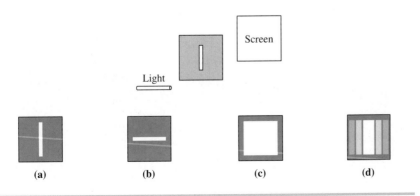

Reflection is an everyday experience.

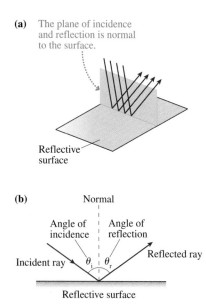

(a) The plane of incidence and reflection is normal to the surface.

Reflective surface

(b) Normal

Angle of incidence   Angle of reflection

Incident ray θ_i θ_r Reflected ray

Reflective surface

FIGURE 23.7 Specular reflection of light.

23.2 Reflection

Reflection of light is a familiar, everyday experience. You see your reflection in the bathroom mirror first thing every morning, reflections in your car's rear-view mirror as you drive to school, and the sky reflected in puddles of standing water. Reflection from a flat, smooth surface, such as a mirror or a piece of polished metal, is called **specular reflection,** from *speculum,* the Latin word for "mirror."

Figure 23.7a shows a bundle of parallel light rays reflecting from a mirror-like surface. You can see that the incident and reflected rays are both in a plane that is normal, or perpendicular, to the reflective surface. A three-dimensional perspective accurately shows the relation between the light rays and the surface, but figures such as this are hard to draw by hand. Instead, it is customary to represent reflection with the simpler pictorial representation of Figure 23.7b. In this figure,

- The plane of the page is the plane of incidence and reflection. The reflective surface extends into and out of the page.
- A *single* light ray represents the entire bundle of parallel rays. This is oversimplified, but it keeps the figure and the analysis clear.

The angle θ_i between the ray and a line perpendicular to the surface—the *normal* to the surface—is called the **angle of incidence.** Similarly, the **angle of reflection** θ_r is the angle between the reflected ray and the normal to the surface. The **law of reflection,** easily demonstrated with simple experiments, states that

1. The incident ray and the reflected ray are in the same plane normal to the surface, and
2. The angle of reflection equals the angle of incidence: $\theta_r = \theta_i$.

NOTE ▶ Optics calculations *always* use the angle measured from the normal, not the angle between the ray and the surface. ◀

EXAMPLE 23.1 Light reflecting from a mirror

A dressing mirror on a closet door is 1.5 m tall. The bottom is 0.5 m above the floor. A bare light bulb hangs 1.0 m from the closet door, 2.5 m above the floor. How long is the streak of reflected light across the floor?

MODEL Treat the light bulb as a point source and use the ray model of light.

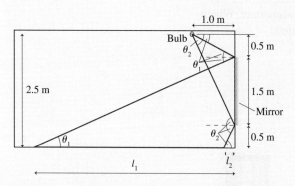

FIGURE 23.8 Pictorial representation of the light rays reflecting from a mirror.

VISUALIZE Figure 23.8 is a pictorial representation of the light rays. We need to consider only the two rays that strike the edges of the mirror. All other reflected rays will fall between these two.

SOLVE Figure 23.8 has used the law of reflection to set the angles of reflection equal to the angles of incidence. Other angles have been identified with simple geometry. The two angles of incidence are

$$\theta_1 = \tan^{-1}\left(\frac{0.5\ \text{m}}{1.0\ \text{m}}\right) = 26.6°$$

$$\theta_2 = \tan^{-1}\left(\frac{2.0\ \text{m}}{1.0\ \text{m}}\right) = 63.4°$$

The distances to the points where the rays strike the floor are then

$$l_1 = \frac{2.0\ \text{m}}{\tan\theta_1} = 4.00\ \text{m}$$

$$l_2 = \frac{0.5\ \text{m}}{\tan\theta_2} = 0.25\ \text{m}$$

Thus the length of the light streak is $l_1 - l_2 = 3.75$ m.

Diffuse Reflection

Most objects are seen by virtue of their reflected light. For a "rough" surface, the law of reflection $\theta_r = \theta_i$ is obeyed at each point but the irregularities of the surface cause the reflected rays to leave in many random directions. This situation, shown in Figure 23.9, is called **diffuse reflection.** It is how you see this page, the wall, your hand, your friend, and so on. Diffuse reflection is far more prevalent than the mirror-like specular reflection.

By a "rough" surface, we mean a surface that is rough or irregular in comparison to the wavelength of light. Because visible light wavelengths are $\approx 0.5\ \mu m$, any surface with texture, scratches, or other irregularities larger than $1\ \mu m$ will cause diffuse reflection rather than specular reflection. A piece of paper may feel quite smooth to your hand, but a microscope would show that the surface consists of distinct fibers much larger than $1\ \mu m$ in size. By contrast, the irregularities on a mirror or a piece of polished metal are much smaller than $1\ \mu m$. The law of reflection is equally valid for both specular and diffuse reflection, but the nature of the surface causes the outcomes to be quite different.

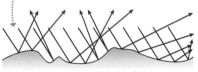

Each ray obeys the law of reflection at that point, but the irregular surface causes the reflected rays to leave in many random directions.

Magnified view of surface

FIGURE 23.9 Diffuse reflection from an irregular surface.

The Plane Mirror

One of the most commonplace observations is that you can see yourself in a mirror. How? Figure 23.10a shows rays from point source P reflecting from a mirror. Consider the particular ray shown in Figure 23.10b. The reflected ray travels along a line that passes through point P′ on the "back side" of the mirror. Because $\theta_r = \theta_i$, simple geometry dictates that P′ is the same distance behind the mirror as P is in front of the mirror. That is, $s' = s$.

Act|v ONLINE Physics   15.4

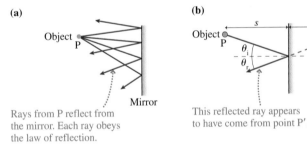

(a) Object P — Rays from P reflect from the mirror. Each ray obeys the law of reflection. — Mirror

(b) Object P — θ_i θ_r — s — s' — P′ — This reflected ray appears to have come from point P′.

(c) Object distance   Image distance — Object P — s — s' — P′ Virtual image — Eye — The reflected rays *all* diverge from P′, which appears to be the source of the reflected rays. Your eye collects the bundle of diverging rays and "sees" the light coming from P′.

FIGURE 23.10 The light rays reflecting from a plane mirror.

The location of point P′ in Figure 23.10b is independent of the value of θ_i. Consequently, *all* reflected rays travel along lines that pass through the *same* point P′. The original light rays diverged from point P, but the reflected rays now diverge from point P′. Consequently, as Figure 23.10c shows, **the reflected rays all *appear* to be coming from point P′.** For a plane mirror, the distance s' to point P′ is equal to the object distance s:

$$s' = s \quad \text{(plane mirror)} \tag{23.3}$$

If rays diverge from an *object point* P and interact with a mirror so that the reflected rays diverge from point P′ and *appear* to come from P′, then we call P′ a **virtual image** of point P. The image is "virtual" in the sense that no rays actually leave P′, which is in darkness behind the mirror. But as far as your eye is concerned, the light rays act exactly *as if* the light really originated at P′. So while you may say "I see P in the mirror," what you are actually seeing is the virtual image of P. Distance s' is the *image distance*.

For an extended object, such as the one in Figure 23.11, each point on the object from which rays strike the mirror has a corresponding image point an equal

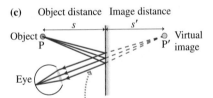

s_P $s_{P'}$ — P — P′ — Q — Q′ — The rays from P and Q that reach your eye reflect from different areas of the mirror.

Your eye intercepts only a very small fraction of all the reflected rays.

FIGURE 23.11 Each point on the extended object has a corresponding image point an equal distance on the opposite side of the mirror.

distance on the opposite side of the mirror. The eye captures and focuses diverging bundles of rays from each point of the image in order to see the full image in the mirror. Two facts are worth noting:

1. Rays from each point on the object spread out in all directions and strike *every point* on the mirror. Only a very few of these rays enter your eye, but the other rays are very real and might be seen by other observers.
2. Rays from points P and Q enter your eye after reflecting from *different* areas of the mirror. This is why you can't always see the full image of an object in a very small mirror.

EXAMPLE 23.2 How high is the mirror?

If your height is h, what is the shortest mirror on the wall in which you can see your full image? Where must the top of the mirror be hung?

MODEL Use the ray model of light.

VISUALIZE Figure 23.12 is a pictorial representation of the light rays. We need to consider only the two rays that leave your head and feet and reflect into your eye.

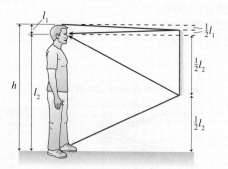

FIGURE 23.12 Pictorial representation of light rays from your head and feet reflecting into your eye.

SOLVE Let the distance from your eyes to the top of your head be l_1 and the distance to your feet be l_2. Your height is $h = l_1 + l_2$. A light ray from the top of your head that reflects from the mirror at $\theta_r = \theta_i$ and enters your eye must, by congruent triangles, strike the mirror a distance $\frac{1}{2}l_1$ above your eyes. Similarly, a ray from your foot to your eye strikes the mirror a distance $\frac{1}{2}l_2$ below your eyes. The distance between these two points on the mirror is $\frac{1}{2}l_1 + \frac{1}{2}l_2 = \frac{1}{2}h$. A ray from anywhere else on your body will reach your eye if it strikes the mirror between these two points. Pieces of the mirror outside these two points are irrelevant, not because rays don't strike them but because the reflected rays don't reach your eye. Thus the shortest mirror in which you can see your full reflection is $\frac{1}{2}h$. But this will only work if the top of the mirror is hung midway between your eyes and the top of your head.

ASSESS It is interesting that the answer does not depend on how far you are from the mirror.

Left and Right

It's common wisdom that a mirror "reverses left and right." But why, then, does it not also reverse up and down? What's special about left and right?

Hold your hands in front of you so that you can see the back of your right hand but the palm of your left hand. If the lighting were poor so that you could see only the outlines of your hands, as in Figure 23.13, could you tell which is a "right hand" and which is a "left hand"? No! Unlike "up" and "down," the terms "right" and "left" do not have an absolute, unambiguous meaning. Right and left are determined by the orientation of your thumb *relative to* your palm. Without knowing where the palm is, you can't assign a handedness to a hand.

In fact, a mirror does not "reverse right and left" any more than it reverses up and down. Instead, a mirror reverses *front and back*. Hold your right hand out, palm away from you and thumb pointing left. Imagine turning your hand inside-out in the sense that everything on the palm side is pulled through toward you and everything on the back side is pushed through away from you. Your fingers would still point up and your thumb would still point to the left, but you would see your palm rather than the back of your hand. Neither up/down nor left/right have reversed, but your "right hand" has become a "left hand."

Right or left?

FIGURE 23.13 Can you tell which hand this is?

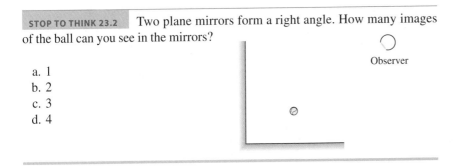

STOP TO THINK 23.2 Two plane mirrors form a right angle. How many images of the ball can you see in the mirrors?

a. 1
b. 2
c. 3
d. 4

Observer

23.3 Refraction

It has been known since antiquity that two things happen when a light ray is incident on a smooth boundary between two transparent materials, such as the boundary between air and glass:

Activ
Physics   15.1–15.3

1. Part of the light *reflects* from the boundary, obeying the law of reflection. This is how you see reflections from pools of water or storefront windows, even though water and glass are transparent.
2. Part of the light continues into the second medium. It is *transmitted* rather than reflected, but the transmitted ray changes direction as it crosses the boundary. The transmission of light from one medium to another, but with a change in direction, is called **refraction.**

The photograph of Figure 23.14 shows the refraction of a laser beam as it passes through a glass prism. Notice that the ray direction changes as the light enters and leaves the glass. You can also see two weak reflections leaving the top surface of the prism.

Reflection from the boundary between transparent media is usually weak. Typically 95% of the light is transmitted and only 5% is reflected. Our goal in this section is to understand refraction, so we will usually ignore the weak reflection and focus on the transmitted light.

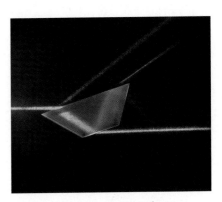

FIGURE 23.14 A laser beam refracts through a glass prism.

NOTE ▶ A transparent material through which light travels is called the *medium* (plural *media*). This term has to be used with caution. The material does affect the light speed, but a transparent material differs from the medium of a sound or water wave in that particles of the medium do *not* oscillate as a light wave passes through. For a light wave it is the electromagnetic field that oscillates. ◀

Figure 23.15a shows the refraction of light rays in a parallel beam of light, such as a laser beam, and rays from a point source. It's good to remember that an infinite number of rays are incident on the boundary, but our analysis will be simplified if we focus on a single light ray. Figure 23.15b is a ray diagram showing

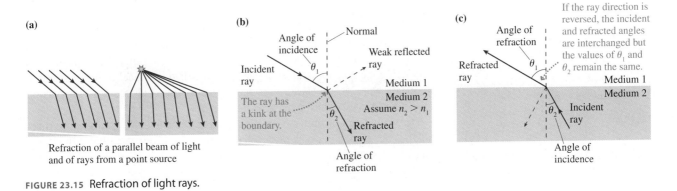

(a)

Refraction of a parallel beam of light and of rays from a point source

(b)

Angle of incidence
Normal
Weak reflected ray
Incident ray
θ_1
Medium 1
Medium 2
Assume $n_2 > n_1$
The ray has a kink at the boundary.
θ_2
Refracted ray
Angle of refraction

(c)

Angle of refraction
If the ray direction is reversed, the incident and refracted angles are interchanged but the values of θ_1 and θ_2 remain the same.
Refracted ray
θ_1
Medium 1
Medium 2
Incident ray
θ_2
Angle of incidence

FIGURE 23.15 Refraction of light rays.

the refraction of a single ray at a boundary between medium 1 and medium 2. Let the angle between the ray and the normal be θ_1 in medium 1 and θ_2 in medium 2. For the medium in which the ray is approaching the boundary, this is the *angle of incidence* as we've previously defined it. The angle on the transmitted side, *measured from the normal,* is called the **angle of refraction.** Notice that θ_1 is the angle of incidence in Figure 23.15b and the angle of refraction in Figure 23.15c, where the ray is traveling in the opposite direction, even though the value of θ_1 has not changed.

The relationship between angles θ_1 and θ_2 is generally believed to have been discovered in 1621 by the Dutch scientist Willebrod Snell, although some historians of science believe it may have been known before then. Regardless of its discoverer, we refer to the "law of refraction" as **Snell's law.** If a ray refracts between medium 1 and medium 2, having indices of refraction n_1 and n_2, the ray angles θ_1 and θ_2 in the two media are related by

$$n_1 \sin\theta_1 = n_2 \sin\theta_2 \quad \text{(Snell's law of refraction)} \tag{23.4}$$

Notice that Snell's law does not mention which is the incident angle and which the refracted angle.

The Index of Refraction

To Snell and his contemporaries, n was simply an "index of the refractive power" of a transparent substance. The relation between the index of refraction and the speed of light was not recognized until the development of a wave theory of light in the 19th century. Theory predicts, and experiment confirms, that light travels through a transparent medium, such as glass or water, at a speed *less* than its speed c in vacuum. We define the *index of refraction n* of a transparent medium as

$$n = \frac{c}{v_{\text{medium}}} \tag{23.5}$$

where v_{medium} is the light speed in the medium. This implies, of course, that $v_{\text{medium}} = c/n$. The index of refraction of a medium is always $n > 1$ except for vacuum, which has $n = 1$ exactly.

Table 23.1 shows measured values of n for several materials. There are many types of glass, each with a slightly different index of refraction, so we will keep things simple by accepting $n = 1.50$ as a typical value. Notice that zircon, the material used to make "cubic zirconium" costume jewelry, has an index of refraction much higher than glass, although not nearly equal to diamond.

We can accept Snell's law as simply an empirical discovery about the behavior of light. Alternatively, and perhaps surprisingly, we can use the wave model of light to justify Snell's law. Figure 23.16 shows the wave front at time t of a plane wave approaching a boundary at speed $v_1 = c/n_1$. The light rays are perpendicular to the wave front, so the angle of incidence θ_1 is also the angle between the wave front and the boundary.

Huygens' principle, which you learned in Chapter 22, says that we can locate the wave front at a later time by thinking of each point on the wave front as the source of spherical wavelets. A spherical wavelet from point B travels distance $v_1\Delta t$ during the time interval Δt. Because point A had just reached the boundary at time t, its wavelet travels through medium 2 at speed $v_2 = c/n_2$. This wavelet travels distance $v_2\Delta t$ during the time interval Δt.

The wave front at time $t + \Delta t$ is tangent to the wavelets. You can see that a kink has developed because of the change in wave speed. The rays in medium 2 are still perpendicular to the wave front, but they now travel at the refracted angle θ_2. Notice that the two right triangles AB′B and AB′A′ share a common

TABLE 23.1 Indices of refraction

Medium	n
Vacuum	1.00 exactly
Air (actual)	1.0003
Air (accepted)	1.00
Water	1.33
Ethyl alcohol	1.36
Oil	1.46
Glass (typical)	1.50
Polystyrene plastic	1.59
Zircon	1.96
Diamond	2.41
Silicon (infrared)	3.50

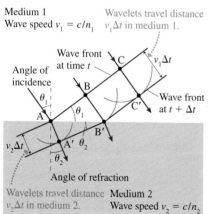

FIGURE 23.16 Snell's law is a consequence of Huygens' principle.

hypotenuse AB'. Calculating the length of the hypotenuse for both triangles, using trigonometry, and equating the results gives

$$\frac{v_1 \Delta t}{\sin \theta_1} = \frac{v_2 \Delta t}{\sin \theta_2} \tag{23.6}$$

The Δt cancels and Equation 23.6 can be written

$$\frac{1}{v_1} \sin \theta_1 = \frac{1}{v_2} \sin \theta_2 \tag{23.7}$$

If we now multiply both sides by c and use the definitions $n_1 = c/v_1$ and $n_2 = c/v_2$, we arrive at

$$n_1 \sin \theta_1 = n_2 \sin \theta_2 \tag{23.8}$$

which is Snell's law.

Examples of Refraction

Look back at Figure 23.15. As the ray in Figure 23.15b moves from medium 1 to medium 2, where $n_2 > n_1$, it bends closer to the normal. In Figure 23.15c, where the ray moves from medium 2 to medium 1, it bends away from the normal. This is a general conclusion that follows from Snell's law:

- When a ray is transmitted into a material with a higher index of refraction, it bends toward the normal.
- When a ray is transmitted into a material with a lower index of refraction, it bends away from the normal.

This rule becomes a central idea in a procedure for analyzing refraction problems.

TACTICS BOX 23.1 Analyzing refraction

❶ **Draw a ray diagram.** Represent the light beam with one ray.
❷ **Draw a line normal to the boundary.** Do this at each point where the ray intersects a boundary.
❸ **Show the ray bending in the correct direction.** The angle is larger on the side with the smaller index of refraction. This is the qualitative application of Snell's law.
❹ **Label angles of incidence and refraction.** Measure all angles from the normal.
❺ **Use Snell's law.** Calculate the unknown angle or unknown index of refraction.

EXAMPLE 23.3 Deflecting a laser beam

A laser beam is aimed at a 1.0-cm-thick sheet of glass at an angle 30° above the glass.

a. What is the laser beam's direction of travel in the glass?
b. What is its direction in the air on the other side?
c. By what distance is the laser beam displaced?

MODEL Represent the laser beam with a single ray and use the ray model of light.

VISUALIZE Figure 23.17 is a pictorial representation in which the first four steps of Tactics Box 23.1 have been identified. Notice that the angle of incidence is $\theta_1 = 60°$, not the 30° value given in the problem.

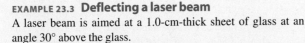

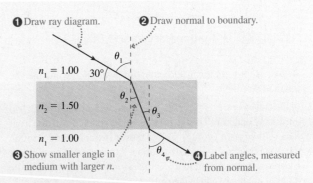

FIGURE 23.17 The ray diagram of a laser beam passing through a sheet of glass.

SOLVE

a. Snell's law, the final step in the Tactics Box, is $n_1 \sin\theta_1 = n_2 \sin\theta_2$. Using $\theta_1 = 60°$, we find that the direction of travel in the glass is

$$\theta_2 = \sin^{-1}\left(\frac{n_1 \sin\theta_1}{n_2}\right) = \sin^{-1}\left(\frac{\sin 60°}{1.5}\right)$$

$$= \sin^{-1}(0.577) = 35.3°$$

b. Snell's law at the second boundary is $n_2 \sin\theta_3 = n_1 \sin\theta_4$. You can see from Figure 23.17 that the interior angles are equal: $\theta_3 = \theta_2 = 35.3°$. Thus the ray emerges back into the air traveling at angle

$$\theta_4 = \sin^{-1}\left(\frac{n_2 \sin\theta_3}{n_1}\right) = \sin^{-1}(1.5 \sin 35.3°)$$

$$= \sin^{-1}(0.867) = 60°$$

This is the same as θ_1, the original angle of incidence. The glass doesn't change the direction of the laser beam.

c. Although the exiting laser beam is parallel to the initial laser beam, it has been displaced sideways by distance d. Figure 23.18 shows the geometry for finding d. From trigonometry, $d = l \sin\phi$. Further $\phi = \theta_1 - \theta_2$ and $l = t/\cos\theta_2$, where t is the thickness of the glass. Combining these gives

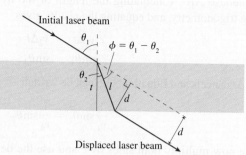

FIGURE 23.18 The laser beam is deflected sideways by distance d.

$$d = l \sin\phi = \frac{t}{\cos\theta_2}\sin(\theta_1 - \theta_2)$$

$$= \frac{(1.0 \text{ cm})\sin 24.7°}{\cos 35.3°} = 0.51 \text{ cm}$$

The glass causes the laser beam to be displaced sideways by 0.51 cm.

ASSESS The laser beam exits the glass still traveling in the same direction as it entered. This is a general result for light traveling through a medium with parallel sides. Notice that the displacement d becomes zero in the limit $t \to 0$. This will be an important observation when we get to lenses.

EXAMPLE 23.4 Measuring the index of refraction

Figure 23.19 shows a laser beam deflected by a 30°-60°-90° prism. What is the prism's index of refraction?

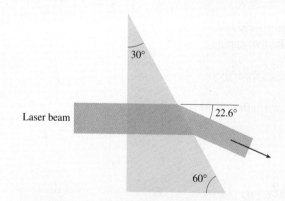

FIGURE 23.19 A prism deflects a laser beam.

MODEL Represent the laser beam with a single ray and use the ray model of light.

VISUALIZE Figure 23.20 uses the steps of Tactics Box 23.1 to draw a ray diagram. The ray is incident perpendicular to the front face of the prism ($\theta_{\text{incident}} = 0°$), thus it is transmitted through the first boundary without deflection. At the second boundary it is especially important to *draw the normal to the surface* at the point of incidence and to *measure angles from the normal*.

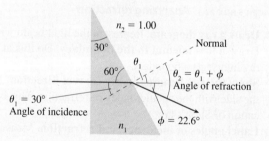

θ_1 and θ_2 are measured from the normal.

FIGURE 23.20 Pictorial representation of a laser beam passing through the prism.

SOLVE From the geometry of the triangle you can find that the laser's angle of incidence on the hypotenuse of the prism is $\theta_1 = 30°$, the same as the apex angle of the prism. The ray exits the prism at angle θ_2 such that the deflection is $\phi = \theta_2 - \theta_1 = 22.6°$. Thus $\theta_2 = 52.6°$. Knowing both angles and $n_2 = 1.00$ for air, we can use Snell's law to find n_1:

$$n_1 = \frac{n_2 \sin\theta_2}{\sin\theta_1} = \frac{1.00 \sin 52.6°}{\sin 30°} = 1.59$$

ASSESS Referring to the indices of refraction in Table 23.1, we see that the prism is made of plastic.

Total Internal Reflection

What would have happened in Example 23.4 if the prism angle had been 45° rather than 30°? The light rays would approach the rear surface of the prism at an angle of incidence $\theta_1 = 45°$. When we try to calculate the angle of refraction at which the ray emerges into the air, we find

$$\sin\theta_2 = \frac{n_1}{n_2}\sin\theta_1 = \frac{1.59}{1.00}\sin 45° = 1.12$$

$$\theta_2 = \sin^{-1}(1.12) = ???$$

Angle θ_2 doesn't compute because the sine of an angle can't be larger than 1. The ray is unable to refract through the boundary. Instead, 100% of the light *reflects* from the boundary back into the prism. This process is called **total internal reflection**, often abbreviated TIR. That it really happens is illustrated in Figure 23.21. Here three laser beams enter a prism from the left. The bottom two refract out through the right side of the prism. The blue beam, which is incident on the prism's back face at a slightly larger angle of incidence, undergoes total internal reflection and then emerges through the right surface.

Figure 23.22 shows several rays leaving a point source in a medium with index of refraction n_1. The medium on the other side of the boundary has $n_2 < n_1$. As we've seen, crossing a boundary into a material with a lower index of refraction causes the ray to bend away from the normal. Two things happen as angle θ_1 increases. First, the refraction angle θ_2 approaches 90°. Second, the fraction of the light energy that is transmitted decreases while the fraction that is reflected increases.

A **critical angle** is reached when $\theta_2 = 90°$. Because $\sin 90° = 1$, Snell's law $n_1 \sin\theta_c = n_2 \sin 90°$ gives the critical angle of incidence as

$$\theta_c = \sin^{-1}\left(\frac{n_2}{n_1}\right) \tag{23.9}$$

The refracted light vanishes at the critical angle and the reflection becomes 100% for any angle $\theta_1 \geq \theta_c$. The critical angle is well defined because of our assumption that $n_2 < n_1$. **There is no critical angle and no total internal reflection if $n_2 > n_1$.**

As a quick example, the critical angle in a typical piece of glass at the glass-air boundary is

$$\theta_{c\ glass} = \sin^{-1}\left(\frac{1.00}{1.50}\right) = 42°$$

The fact that the critical angle is less than 45° has important applications. For example, Figure 23.23 shows a pair of binoculars. The lenses are much farther apart than your eyes, so the light rays need to be brought together before exiting the eyepieces. Rather than using mirrors, which get dirty, are easily scratched, and require alignment, binoculars use a pair of prisms on each side. Thus the light undergoes two total internal reflections and emerges from the eyepiece.

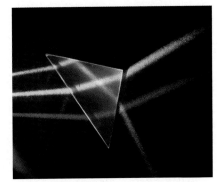

FIGURE 23.21 One of the three laser beams undergoes total internal reflection.

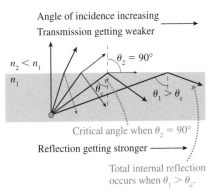

FIGURE 23.22 Refraction and reflection of rays as the angle of incidence increases.

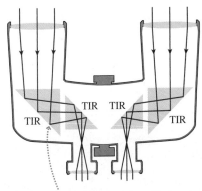

FIGURE 23.23 Binoculars and other optical instruments make use of total internal reflection.

EXAMPLE 23.5 **Total internal reflection**

A light bulb is set in the bottom of a 3.0-m-deep swimming pool. What is the diameter of the circle of light seen on the water's surface from above?

MODEL Represent the light bulb as a point source and use the ray model of light.

VISUALIZE Figure 23.24 on the next page is a pictorial representation of the light rays. The light bulb emits rays at all angles,

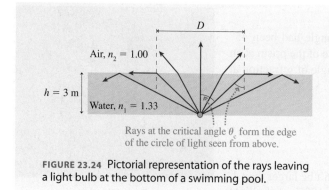

FIGURE 23.24 Pictorial representation of the rays leaving a light bulb at the bottom of a swimming pool.

Rays at the critical angle θ_c form the edge of the circle of light seen from above.

but only some of the rays refract into the air where they can be seen from above. Rays striking the surface at greater than the critical angle undergo TIR and remain within the water. The diameter of the circle of light is the distance between the two points at which rays strike the surface at the critical angle.

SOLVE From trigonometry, the circle diameter is $D = 2h \tan \theta_c$, where h is the depth of the water. The critical angle for a water-air boundary is $\theta_c = \sin^{-1}(1.00/1.33) = 48.7°$. Thus

$$D = 2(3.0 \text{ m}) \tan 48.7° = 6.83 \text{ m}$$

Fiber Optics

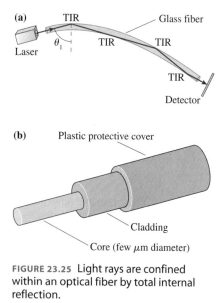

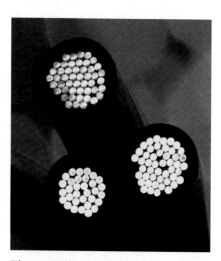

FIGURE 23.25 Light rays are confined within an optical fiber by total internal reflection.

The most important modern application of total internal reflection is the transmission of light through optical fibers. Figure 23.25a shows a laser beam shining into the end of a long, narrow-diameter glass tube. The light rays pass easily from the air into the glass, but they then impinge on the inside wall of the glass tube at an angle of incidence θ_1 approaching 90°. This is well above the critical angle, so the laser beam undergoes TIR and remains inside the glass. The laser beam continues to "bounce" its way down the tube as if the light were inside a pipe. Indeed, optical fibers are sometimes called "light pipes." The rays are *below* the critical angle ($\theta_1 \approx 0$) when they finally reach the end of the fiber, thus they refract out without difficulty and can be detected.

While a simple glass tube can transmit light, reliance on a glass-air boundary is not sufficiently reliable for commercial use. Any small scratch on the side of the tube alters the rays' angle of incidence and allows leakage of light. Figure 23.25b shows the construction of a practical optical fiber. A small-diameter glass *core* is surrounded by a layer of glass *cladding*. The glasses used for the core and the cladding have $n_{\text{core}} > n_{\text{cladding}}$, thus light undergoes TIR at the core-cladding boundary and remains confined within the core. This boundary is not exposed to the environment and hence retains its integrity even under adverse conditions.

Even glass of the highest purity is not perfectly transparent. Absorption in the glass, even if very small, causes a gradual decrease in light intensity. The glass used for the core of optical fibers has a minimum absorption at a wavelength of 1.3 μm, in the infrared, so this is the laser wavelength used for long-distance signal transmission. Light at this wavelength can travel hundreds of kilometers through a fiber without significant loss.

STOP TO THINK 23.3 A light ray travels from medium 1 to medium 3 as shown. For these media,

a. $n_3 > n_1$.
b. $n_3 = n_1$.
c. $n_3 < n_1$.
d. We can't compare n_1 to n_3 without knowing n_2.

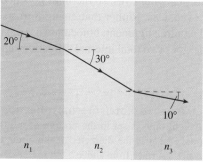

Fiber optics have replaced copper wires for carrying digital signals.

23.4 Image Formation by Refraction

You've likely made an interesting observation while looking at fish in an aquarium. First you see a fish that appears to be swimming close to the front window of the aquarium. But then, if you look through the side of the aquarium, you find that the fish is actually farther from the front window than you thought. Somehow, as you look through the front window, the fish appears to be closer than it really is. This is a puzzle crying out for an explanation.

To begin, recall how vision functions. A diverging bundle of rays leaves the object, enters the pupil of the eye, and is focused on the retina. By adjusting the eye's lens to achieve a good focus, your brain determines the distance d from which the rays originated. This is where you perceive the object to be. Figure 23.26a shows how you would see a fish out of water at distance d.

Now place the fish back in the aquarium at the same distance d. For simplicity, we'll ignore the glass wall of the aquarium and consider the water-air boundary. (The thin glass of a typical window has only a very small effect on the refraction of the rays and doesn't change the conclusions.) Light rays again leave the fish, but this time they refract at the water-air boundary. Because they're going from a higher to a lower index of refraction, the rays refract *away from* the normal. Figure 23.26b shows the consequences.

A bundle of diverging rays still enters your eye, but now these rays *seem* to be diverging from a closer point, at distance d'. As far as your eye and brain are concerned, it's exactly *as if* the rays really originate at distance d', and this is the location at which you "see" the fish. **The object appears closer than it really is because of the refraction of light at the boundary.**

We found that the rays reflected from a mirror diverge from a point that is not the object point. We called that point a *virtual image.* Similarly, if rays from an object point P refract at a boundary between two media such that the rays then diverge from a point P′ and *appear* to come from P′, we call P′ a virtual image of point P. The virtual image of the fish is what you see.

Let's examine this image formation a bit more carefully. Figure 23.27 shows a boundary between two transparent media having indices of refraction n_1 and n_2. Point P, a source of light rays, is the object. Point P′, from which the rays *appear* to diverge, is the virtual image of P. The figure assumes $n_1 > n_2$, but this assumption isn't necessary. Distance s is called the **object distance.** Our goal is to determine distance s', the **image distance.**

A line perpendicular to the boundary is called the **optical axis.** Consider a ray that leaves the object at angle θ_1 with respect to the optical axis. θ_1 is also the angle of incidence at the boundary, where the ray refracts into the second medium at angle θ_2. By tracing the refracted ray backward, you can see that θ_2 is also the angle between the refracted ray and the optical axis at point P′.

The distance l is common to both the incident and the refracted rays, and you can see that $l = s \tan\theta_1 = s' \tan\theta_2$. Thus

$$s' = \frac{\tan\theta_1}{\tan\theta_2} s \tag{23.10}$$

Snell's law relates the sines of angles θ_1 and θ_2. That is,

$$\frac{\sin\theta_1}{\sin\theta_2} = \frac{n_2}{n_1} \tag{23.11}$$

In practice, the angle between any of these rays and the optical axis is very small because the size of the pupil of your eye is very much less than the distance between the object and your eye. (The angles in the figure have been greatly exaggerated.) Rays that are nearly *parallel* to the *axis* are called **paraxial rays.**

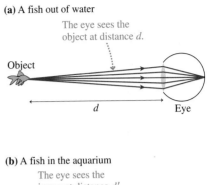

(a) A fish out of water

The eye sees the object at distance d.

Object

d Eye

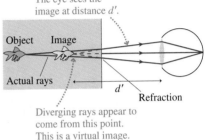

(b) A fish in the aquarium

The eye sees the image at distance d'.

Object   Image

Actual rays

d' Refraction

Diverging rays appear to come from this point. This is a virtual image.

FIGURE 23.26 Refraction of the light rays causes a fish in the aquarium to be seen at distance d'.

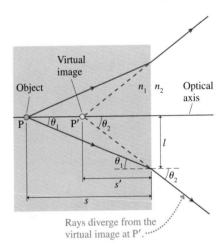

Virtual image

Object

n_1 n_2 Optical axis

P θ_1 P′ θ_2

θ_1

l

θ_2

s'

s

Rays diverge from the virtual image at P′.

FIGURE 23.27 Finding the virtual image P′ of an object at P.

The small-angle approximation $\sin\theta \approx \tan\theta \approx \theta$, where θ is in radians, can be applied to paraxial rays. Consequently,

$$\frac{\tan\theta_1}{\tan\theta_2} \approx \frac{\sin\theta_1}{\sin\theta_2} = \frac{n_2}{n_1} \tag{23.12}$$

Using this result in Equation 23.10, we find that the image distance is

$$s' = \frac{n_2}{n_1}s \tag{23.13}$$

NOTE ▶ The fact that the result for s' is independent of θ_1 implies that *all* paraxial rays appear to diverge from the same point P'. This property of the diverging rays is essential in order to have a well-defined image. ◀

This section has given us a first look at image formation via refraction. We will extend this idea to image formation with lenses in Section 23.6.

EXAMPLE 23.6 An air bubble in a window
A fish and a sailor look at each other through a 5.0-cm-thick glass porthole in a submarine. There happens to be a small air bubble right in the center of the glass. How far behind the surface of the glass does the air bubble appear to the fish? To the sailor?

MODEL Represent the air bubble as a point source and use the ray model of light.

VISUALIZE Paraxial light rays from the bubble refract into the air on one side and into the water on the other. The ray diagram looks like Figure 23.27.

SOLVE The index of refraction of the glass is $n_1 = 1.50$. The bubble is in the center of the window, so the object distance

from either side of the window is $s = 2.5$ cm. From the water side, the fish sees the bubble at an image distance

$$s' = \frac{n_2}{n_1}s = \frac{1.33}{1.50}(2.5 \text{ cm}) = 2.2 \text{ cm}$$

The sailor, in air, sees the bubble at an image distance

$$s' = \frac{n_2}{n_1}s = \frac{1.00}{1.50}(2.5 \text{ cm}) = 1.7 \text{ cm}$$

ASSESS The image distance is *less* for the sailor because of the *larger* difference between the two indices of refraction.

23.5 Color and Dispersion

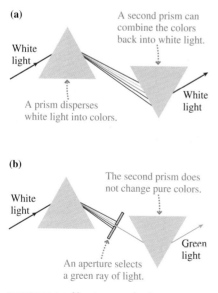

(a)
White light
A prism disperses white light into colors.
A second prism can combine the colors back into white light.
White light

(b)
White light
An aperture selects a green ray of light.
The second prism does not change pure colors.
Green light

FIGURE 23.28 Newton used prisms to study color.

One of the most obvious visual aspects of light is the phenomenon of color. Yet color, for all its vivid sensation, is not inherent in the light itself. Color is a *perception,* not a physical quantity. Color is associated with the wavelength of light, but the fact that we see light with a wavelength of 650 nm as "red" tells us how our visual system responds to electromagnetic waves of this wavelength. There is no "redness" associated with the light wave itself.

Most of the results of optics do not depend on color. Rays of red light pass through glass the same as rays of blue light, so we generally don't need to know the color of light—or, to be more precise, its wavelength—to use the laws of reflection and refraction. Nonetheless, color is an interesting subject, one worthy of a short digression.

Color

It has been known since antiquity that irregularly shaped glass and crystals cause sunlight to be broken into various colors. A common idea was that the glass or crystal somehow altered the properties of the light by *adding* color to the light. Newton suggested a different explanation. He first passed a sunbeam through a prism, producing the familiar rainbow of light. We say that the prism *disperses* the light. Newton's novel idea, shown in Figure 23.28a, was to use a second prism, inverted with respect to the first, to "reassemble" the colors. He found that the light emerging from the second prism was a beam of pure, white light.

But the emerging light beam is white only if *all* the rays are allowed to move between the two prisms. Blocking some of the rays with small obstacles, as in Figure 23.28b, causes the emerging light beam to have color. This suggests that color is associated with the light itself, not with anything that the prism is "doing" to the light. Newton tested this idea by inserting a small aperture between the prisms to pass only the rays of a particular color, such as green. If the prism alters the properties of light, then the second prism should change the green light to other colors. Instead, the light emerging from the second prism is unchanged from the green light entering the prism.

These and similar experiments show that

1. What we perceive as white light is a mixture of all colors. White light can be dispersed into its various colors and, equally important, mixing all the colors produces white light.
2. The index of refraction of a transparent material differs slightly for different colors of light. Glass has a slightly larger index of refraction for violet light than for green light or red light. Consequently, different colors of light refract at slightly different angles and follow slightly different paths through a piece of glass. A prism does not alter the light or add anything to the light; it simply causes the different colors that are inherent in white light to follow slightly different trajectories.

Dispersion

It was Thomas Young, with his two-slit interference experiment, who showed that what we perceive as different colors are associated with light of different wavelengths. The longest wavelengths are perceived as red light and the shortest wavelengths are perceived as violet light. Table 23.2 is a brief summary of the *visible spectrum* of light. Visible-light wavelengths are used so frequently that it is well worth committing this short table to memory.

Newton's observation that the index of refraction varies slightly with color implies that the index of refraction varies slightly with wavelength. This is known as **dispersion.** Figure 23.29 shows the *dispersion curves* of two common glasses. Notice that *n* is *larger* when the wavelength is *shorter,* thus violet light refracts more than red light.

I procured me a triangular glass prism to try therewith the celebrated phenomena of colors.

Isaac Newton

TABLE 23.2 A brief summary of the visible spectrum of light

Color	Approximate wavelength
Deepest red	700 nm
Red	650 nm
Green	550 nm
Blue	450 nm
Deepest violet	400 nm

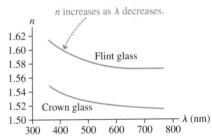

FIGURE 23.29 Dispersion curves show how the index of refraction varies with wavelength.

EXAMPLE 23.7 Dispersing light with a prism

Example 23.4 found that a ray incident on a 30° prism is deflected by 22.6° if the prism's index of refraction is 1.59. Suppose this is the index of refraction of deep violet light, and that deep red light has an index of refraction of 1.54.

a. What is the deflection angle for deep red light?
b. If a beam of white light is dispersed by this prism, how wide is the rainbow spectrum on a screen 2.0 m away?

VISUALIZE Figure 23.20 showed the geometry. A ray of any wavelength is incident on the rear surface of the prism at $\theta_1 = 30°$.

SOLVE

a. If $n_1 = 1.54$ for deep red light, the refraction angle is

$$\theta_2 = \sin^{-1}\left(\frac{n_1 \sin\theta_1}{n_2}\right) = \sin^{-1}\left(\frac{1.54 \sin 30°}{1.00}\right) = 50.4°$$

Example 23.4 showed that the deflection angle is $\phi = \theta_2 - \theta_1$, so deep red light is deflected by $\phi_{red} = 20.4°$. This angle is slightly smaller than the previously observed $\phi_{violet} = 22.6°$.

b. The entire spectrum is spread between $\phi_{red} = 20.4°$ and $\phi_{violet} = 22.6°$. The angular spread is

$$\delta = \phi_{violet} - \phi_{red} = 2.2° = 0.038 \text{ rad}$$

At distance r, the spectrum spans an arc length

$$s = r\delta = (2.0 \text{ m})(0.038 \text{ rad}) = 0.076 \text{ m} = 7.6 \text{ cm}$$

ASSESS The angle is so small that there's no appreciable difference between arc length and a straight line. The spectrum will be 7.6 cm wide at a distance of 2.0 m.

Rainbows

One of the most interesting sources of color in nature is the rainbow. The details get somewhat complicated, but Figure 23.30a shows that the basic cause of the rainbow is a combination of refraction, reflection, and dispersion.

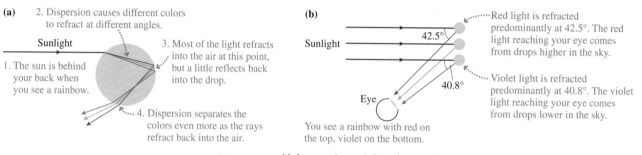

(a)
2. Dispersion causes different colors to refract at different angles.

Sunlight

1. The sun is behind your back when you see a rainbow.

3. Most of the light refracts into the air at this point, but a little reflects back into the drop.

4. Dispersion separates the colors even more as the rays refract back into the air.

(b)
Sunlight

42.5°

40.8°

Eye

Red light is refracted predominantly at 42.5°. The red light reaching your eye comes from drops higher in the sky.

Violet light is refracted predominantly at 40.8°. The violet light reaching your eye comes from drops lower in the sky.

You see a rainbow with red on the top, violet on the bottom.

FIGURE 23.30 Light seen in a rainbow has undergone refraction + reflection + refraction in a raindrop.

Figure 23.30a might lead you to think that the top edge of a rainbow is violet. In fact, the top edge is red, and violet is on the bottom. The rays leaving the drop in Figure 23.30a are spreading apart, so they can't all reach your eye. As Figure 23.30b shows, a ray of red light reaching your eye comes from a drop *higher* in the sky than a ray of violet light. In other words, the colors you see in a rainbow refract toward your eye from different raindrops, not from the same drop. You have to look higher in the sky to see the red light than to see the violet light.

Colored Filters and Colored Objects

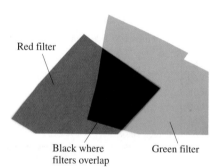

Red filter

Black where filters overlap

Green filter

No light at all passes through both a green and a red filter.

White light passing through a piece of green glass emerges as green light. A possible explanation would be that the green glass *adds* "greenness" to the white light, but Newton found otherwise. Green glass is green because it *removes* any light that is "not green." More precisely, a piece of colored glass *absorbs* all wavelengths except those of one color, and that color is transmitted through the glass without hindrance. We can think of a piece of colored glass or plastic as a *filter* that removes all wavelengths except a chosen few.

EXAMPLE 23.8 Filtering light

White light passes through a green filter and is observed on a screen. Describe how the screen will look if a second green filter is placed between the first filter and the screen. Describe how the screen will look if a red filter is placed between the green filter and the screen.

VISUALIZE The first filter removes all light except for wavelengths near 550 nm that we perceive as green light. A second green filter doesn't have anything to do. The nongreen wavelengths have already been removed, and the green light emerging from the first filter will pass through the second filter without difficulty. The screen will continue to be green and its intensity will not change. A red filter, by contrast, absorbs all wavelengths except those near 650 nm. The red filter will absorb the green light, and *no* light will reach the screen. The screen will be dark.

This behavior is true not just for glass filters, which transmit light, but for *pigments* that absorb light of some wavelengths but *reflect* light at other wavelengths. For example, red paint contains pigments that reflect light at wavelengths near 650 nm while absorbing all other wavelengths. Pigments in paints, inks, and natural objects are responsible for most of the color we observe in the world, from the red of lipstick to the blue of a bluebird's feathers.

As an example, Figure 23.31 shows the absorption curve of *chlorophyll*. Chlorophyll is essential for photosynthesis in green plants. The chemical reactions of photosynthesis are able to use red light and blue/violet light, thus chlorophyll has evolved to absorb red light and blue/violet light from sunlight and put it to use. But green and yellow light are not absorbed. Instead, to conserve energy, these wavelengths are mostly *reflected* to give the object a greenish-yellow color. When you look at the green leaves on a tree, you're seeing the light that was reflected because it *wasn't* needed for photosynthesis.

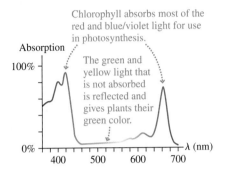

FIGURE 23.31 The absorption curve of chlorophyll.

Light Scattering: Blue Skies and Red Sunsets

In the ray model of Section 23.1 we noted that light within a medium can be scattered or absorbed. As we've now seen, the absorption of light can be wavelength dependent and can create color in objects. What are the effects of scattering?

Light can scatter from small particles that are suspended in a medium. If the particles are large compared to the wavelengths of light—even though they may be microscopic and not readily visible to the naked eye—the light essentially reflects off the particles. The law of reflection doesn't depend on wavelength, so all colors are scattered equally. White light scattered from many small particles makes the medium appear cloudy and white. Two well-known examples are clouds, where micrometer-size water droplets scatter the light, and milk, which is a colloidal suspension of microscopic droplets of fats and proteins.

A more interesting aspect of scattering occurs at the atomic level. The atoms and molecules of a transparent medium are much smaller than the wavelengths of light, so they can't scatter light simply by reflection. Instead, the oscillating electric field of the light wave interacts with the electrons in each atom in such a way that the light is scattered. This atomic-level scattering is called **Rayleigh scattering.**

Unlike the scattering by small particles, Rayleigh scattering from atoms and molecules *does* depend on the wavelength. A detailed analysis shows that the intensity of scattered light depends inversely on the fourth power of the wavelength: $I_{scattered} \propto \lambda^{-4}$. This wavelength dependence explains why the sky is blue and sunsets are red.

As sunlight travels through the atmosphere, the λ^{-4} dependence of Rayleigh scattering causes the shorter wavelengths to be preferentially scattered. If we take 650 nm as a typical wavelength for red light and 450 nm for blue light, the intensity of scattered blue light relative to scattered red light is

$$\frac{I_{blue}}{I_{red}} = \left(\frac{650}{450}\right)^4 \approx 4$$

Sunsets are red because all the blue light has scattered as the sunlight passes through the atmosphere.

Four times more blue light is scattered toward us than red light and thus, as Figure 23.32 shows, the sky appears blue.

Because of the earth's curvature, sunlight has to travel much farther through the atmosphere when we see it at sunrise or sunset than it does during the midday hours. In fact, the path length through the atmosphere at sunset is so long that essentially all the short wavelengths have been lost due to Rayleigh scattering. Only the longer wavelengths remain—orange and red—and they make the colors of the sunset.

23.6 Thin Lenses: Ray Tracing

A camera obscura or a pinhole camera forms images on a screen, but the images are faint and not perfectly focused. The ability to create a bright, well-focused image is vastly improved by using a lens. A **lens** is a transparent material that uses refraction at *curved* surfaces to form an image from diverging light rays. We will

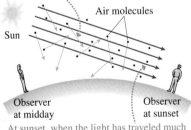

At midday the scattered light is mostly blue because molecules preferentially scatter shorter wavelengths.

At sunset, when the light has traveled much farther through the atmosphere, the light is mostly red because the shorter wavelengths have been lost to scattering.

FIGURE 23.32 Rayleigh scattering by molecules in the air gives the sky and sunsets their color.

defer a detailed analysis of the refraction of lenses until the next section. First, we want to establish a pictorial method of understanding image formation. This method is called **ray tracing.**

Figure 23.33 shows parallel light rays entering two different lenses. The left lens, called a **converging lens,** causes the rays to refract *toward* the optical axis. The common point through which initially parallel rays pass is called the **focal point** of the lens. The distance of the focal point from the lens is called the **focal length** f of the lens. The right lens, called a **diverging lens,** refracts parallel rays *away from* the optical axis. This lens also has a focal point, but it is not as obvious in the figure.

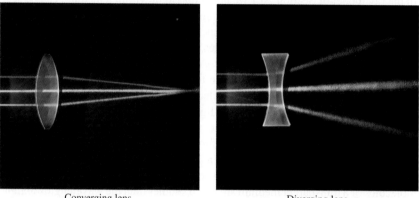

Converging lens   Diverging lens

FIGURE 23.33 Parallel light rays pass through a converging lens and a diverging lens.

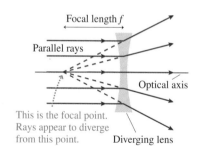

FIGURE 23.34 The focal point and focal length of converging and diverging lenses.

NOTE ▶ A converging lens is thicker in the center than at the edges. A diverging lens is thicker at the edges than at the center. ◀

Figure 23.34 clarifies the situation. In the case of a diverging lens, a backward projection of the diverging rays shows that they all *appear* to have started from the same point. This is the focal point of a diverging lens, and its distance from the lens is the focal length of the lens. In the next section we'll relate the focal length to the curvature and index of refraction of the lens, but for now we'll use the practical definition that **the focal length is the distance from the lens at which rays parallel to the optical axis converge or from which they diverge.**

NOTE ▶ The focal length f is a property *of the lens,* independent of how the lens is used. The focal length characterizes a lens in much the same way that a mass m characterizes an object or a spring constant k characterizes a spring. ◀

Converging Lenses

These basic observations about lenses are enough to understand image formation by a thin lens. A **thin lens** is a lens whose thickness is very small in comparison to its focal length and in comparison to the object and image distances. We'll make the approximation that the thickness of a thin lens is zero and that the lens lies in a plane called the **lens plane.** Within this approximation, all refraction occurs as the rays cross the lens plane, and all distances are measured from the lens plane. Fortunately, the thin-lens approximation is quite good for most practical applications of lenses.

NOTE ▶ We'll *draw* lenses as if they have a thickness, because that is how we expect lenses to look, but our analysis will not depend on the shape or thickness of a lens. ◀

Figure 23.35 shows three important situations of light rays passing through a thin, converging lens. Part a is familiar from Figure 23.34. If the direction of each of the rays in Figure 23.35a is reversed, Snell's law tells us that each ray will exactly retrace its path and emerge from the lens parallel to the optical axis. This leads to Figure 23.35b, which is the "mirror image" of part (a).

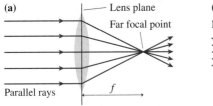

(a)

Any ray initially parallel to the optical axis will refract through the focal point on the far side of the lens.

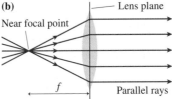

(b)

Any ray passing through the near focal point emerges from the lens parallel to the optical axis.

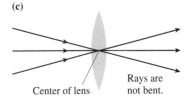

(c)

Any ray directed at the center of the lens passes through in a straight line.

FIGURE 23.35 Three important sets of rays passing through a thin, converging lens.

Notice that the lens actually has *two* focal points, located at distances f on either side of the lens. The focal point on the side from which the light rays are approaching is the *near focal point*. The focal point opposite the side from which the light rays are approaching is the *far focal point*.

Figure 23.35c shows several rays passing through the *center* of the lens. At the center, the two sides of a lens are very nearly parallel to each other. Earlier, in Example 23.3, we found that a ray passing through a piece of glass with parallel sides is *displaced* but *not bent* and that the displacement becomes zero as the thickness approaches zero. Consequently, a ray through the center of a thin lens, with zero thickness, is neither bent nor displaced but travels in a straight line.

These three situations form the basis for ray tracing.

Real Images

Figure 23.36 shows a lens and an object whose distance from the lens is larger than the focal length. Rays from point P on the object are refracted by the lens so as to converge at point P′ on the opposite side of the lens. If rays diverge from an object point P and interact with a lens such that the refracted rays *converge* at point P′, then we call P′ a **real image** of point P. Contrast this with our prior definition of a *virtual image* as a point from which rays *diverge*.

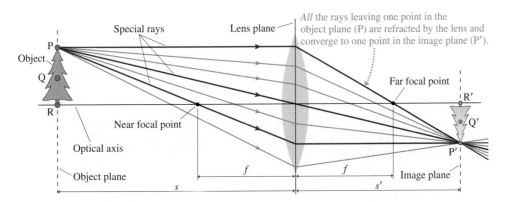

FIGURE 23.36 Rays from an object point P are refracted by the lens and converge to a real image at point P′.

All points on the object that are in the same plane, the **object plane,** converge to image points in the **image plane.** Points Q and R in the object plane of Figure 23.36 have image points Q′ and R′ in the same plane as point P′. Once we locate *one* point in the image plane, such as point P′, we know that the full image lies in the same plane.

There are two important observations to make about Figure 23.36. First, the image is upside down with respect to the object. This is called an **inverted image,** and it is a standard characteristic of real image formation with a converging lens. You have to put slides into a slide tray upside down so that the image seen on the screen is right side up. Second, rays from point P *fill* the entire lens surface, and all portions of the lens contribute to the image. A larger lens will "collect" more rays and thus make a brighter image. This is the big advantage of a lens over a camera obscura or a pinhole camera.

Figure 23.37 is a close-up view of the rays very near the image plane. The rays don't stop at P′ unless we place a screen in the image plane. When we do so, we see a sharp, well-focused image on the screen. To focus an image, you must either move the screen to coincide with the image plane or move the lens or object to make the image plane coincide with the screen. For example, the focus knob on a slide projector moves the lens closer to or farther from the slide until the image plane matches the screen position.

NOTE ▶ The ability to view and record *real* images, where the rays actually converge, sets real images apart from *virtual* images. But keep in mind that we need not *see* a real image in order to *have* an image. A real image exists at a point in space where the rays converge even if there's no viewing screen in the image plane. ◀

Figure 23.36 highlights three "special rays" that are based on the three situations of Figure 23.35. Notice that these three rays alone are sufficient to locate the image point P′. That is, we don't need to draw all the rays shown in Figure 23.36. The procedure known as *ray tracing* consists of locating the image by the use of just these three rays.

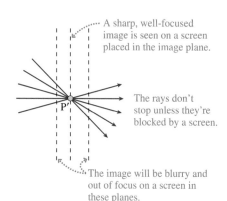

A sharp, well-focused image is seen on a screen placed in the image plane.

The rays don't stop unless they're blocked by a screen.

The image will be blurry and out of focus on a screen in these planes.

FIGURE 23.37 A close-up look at the rays near the image plane.

15.9 Activ Physics ONLINE

TACTICS BOX 23.2 **Ray tracing for a converging lens**

❶ **Draw an optical axis.** Use graph paper or a ruler! Establish an appropriate scale.

❷ **Center the lens on the axis.** Mark and label the focal points at distance f on either side.

❸ **Represent the object with an upright arrow at distance s.** It's usually best to place the base of the arrow on the axis and to draw the arrow about half the radius of the lens.

❹ **Draw the three "special rays" from the tip of the arrow.** Use a straightedge.

    a. A ray parallel to the axis refracts through the far focal point.

    b. A ray that enters the lens along a line through the near focal point emerges parallel to the axis.

    c. A ray through the center of the lens does not bend.

❺ **Extend the rays until they converge.** This is the image point. Draw the rest of the image in the image plane. If the base of the object is on the axis, then the base of the image will also be on the axis.

❻ **Measure the image distance s'.** Also, if needed, measure the image height relative to the object height.

EXAMPLE 23.9 Finding the image of a flower

A 4.0-cm-diameter flower is 200 cm from the 50-cm-focal-length lens of a camera. How far should the film be placed behind the lens to record a well-focused image? What is the diameter of the image on the film?

MODEL The flower is in the object plane. Use ray tracing to locate the image.

VISUALIZE Figure 23.38 shows the ray-tracing diagram and the steps of Tactics Box 23.2. The image has been drawn in the plane where the three special rays converge. You can see *from the drawing* that the image distance is $s' \approx 67$ cm. This is where the film needs to be placed to record a focused image.

The heights of the object and image are labeled h and h'. The ray through the center of the lens is a straight line, thus the object and image both subtend the same angle θ. Using similar triangles,

$$\frac{h'}{s'} = \frac{h}{s}$$

Solving for h' gives

$$h' = h\frac{s'}{s} = (4.0 \text{ cm})\frac{67 \text{ cm}}{200 \text{ cm}} = 1.3 \text{ cm}$$

The flower's image has a diameter of 1.3 cm.

ASSESS We've been able to learn a great deal about the image from a simple geometric procedure.

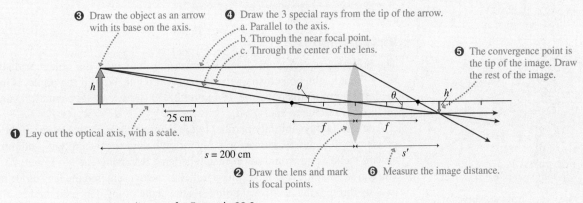

❸ Draw the object as an arrow with its base on the axis.

❹ Draw the 3 special rays from the tip of the arrow.
 a. Parallel to the axis.
 b. Through the near focal point.
 c. Through the center of the lens.

❺ The convergence point is the tip of the image. Draw the rest of the image.

❶ Lay out the optical axis, with a scale.

25 cm

$s = 200$ cm

❷ Draw the lens and mark its focal points.

❻ Measure the image distance.

FIGURE 23.38 Ray-tracing diagram for Example 23.9.

Magnification

The image can be either larger or smaller than the object, depending on the location and focal length of the lens. But there's more to a description of the image than just its size. We also want to know its *orientation* relative to the object. That is, is the image upright or inverted?

Earlier in the chapter, we defined the magnification m as the ratio of image height to object height. It's now useful to expand that definition to include information about the orientation of the image. The revised definition of the **magnification,** which we'll call M, is

$$M = -\frac{h'}{h} = -\frac{s'}{s} \tag{23.14}$$

You just saw in Example 23.9 that the image-to-object height ratio is $h'/h = s'/s$. Consequently, we interpret the magnification M as follows:

1. A positive value of M indicates that the image is upright relative to the object. A negative value of M indicates that the image is inverted relative to the object.
2. The absolute value of M gives the size ratio of the image and object: $h'/h = |M|$.

The magnification in Example 23.9 would be $M = -0.33$, indicating that the image is inverted and 33% the size of the object.

NOTE ▶ "Magnification" can be less than 1, meaning that the image is smaller than the object (i.e., "demagnified"). ◀

A lens produces a sharply focused, inverted image on a screen. What will you see on the screen if the lens is removed?

a. The image will be inverted and blurry.
b. The image will be upright and sharp.
c. The image will be upright and blurry.
d. The image will be much dimmer but otherwise unchanged.
e. There will be no image at all.

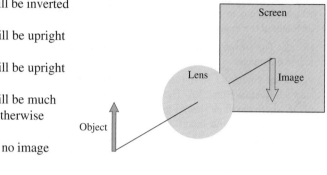

A ray *along a line* through the near focal point refracts parallel to the optical axis.

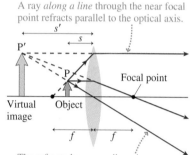

The refracted rays are diverging. They appear to come from point P′.

FIGURE 23.39 Rays from an object at distance $s < f$ are refracted by the lens and diverge to form a virtual image at point P′.

(a)

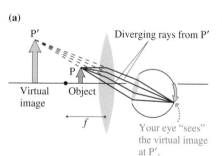

Diverging rays from P′

Your eye "sees" the virtual image at P′.

(b)

FIGURE 23.40 A converging lens is a magnifying glass when the object distance is less than f.

Virtual Images

The previous section considered a converging lens with the object at distance $s > f$. That is, the object was outside the focal point. What if the object is inside the focal point, at distance $s < f$? Figure 23.39 shows just this situation, and we can use ray tracing to analyze it.

The special rays initially parallel to the axis and through the center of the lens present no difficulties. However, a ray through the near focal point would travel toward the left and would never reach the lens! Referring back to Figure 23.35b, you can see that the rays emerging parallel to the axis entered the lens *along a line* passing through the near focal point. It's the angle of incidence on the lens that is important, not whether the light ray actually passes through the focal point. This was the basis for the wording of step 4b in Tactics Box 23.2 and is the third special ray shown in Figure 23.39.

You can see that the three refracted rays don't converge. Instead, all three rays appear to *diverge* from point P′. This is the situation we found for rays reflecting from a mirror and for the rays refracting out of an aquarium. Point P′ is a *virtual image* of the object point P. Furthermore, it is an **upright image,** having the same orientation as the object.

The refracted rays, which are all to the right of the lens, *appear* to come from P′, but none of the rays were ever at that point. No image would appear on a screen placed in the image plane at P′. So what good is a virtual image?

Your eye collects and focuses bundles of diverging rays, thus, as Figure 23.40a shows, you can "see" a virtual image by looking *through* the lens. This is exactly what you do with a magnifying glass, producing a scene like the one in Figure 23.40b. In fact, you view a virtual image anytime you look *through* the eyepiece of an optical instrument such as a microscope or binoculars.

The image distance s' for a virtual image is defined to be a *negative number* $(s' < 0)$, indicating that the image is on the opposite side of the lens from a real image. With this choice of sign, the definition of magnification, $M = -s'/s$, is still valid. A virtual image with negative s' has $M > 0$, implying that the image is upright. This agrees with the ray tracing in Figure 23.39 and the photograph of Figure 23.40b.

NOTE ▶ A lens thicker in the middle than at the edges is classified as a converging lens. The light rays from an object *can* converge to form a real image after passing through such a lens, but only if the object distance is larger than the focal length of the lens: $s > f$. If $s < f$, the rays leaving a converging lens are diverging to produce a virtual image. ◀

EXAMPLE 23.10 Magnifying a flower

To see a flower better, a naturalist holds a 6.0-cm-focal-length magnifying glass 4.0 cm from the flower. What is the magnification?

MODEL The flower is in the object plane. Use ray tracing to locate the image.

VISUALIZE Figure 23.41 shows the ray-tracing diagram. The three special rays diverge from the lens, but we can use a straightedge to extend the rays backward to the point from which they diverge. This point, the image point, is seen to be 12 cm to the left of the lens. Because this is a virtual image, the image distance is $s' = -12$ cm. Thus the magnification is

$$M = -\frac{s'}{s} = -\frac{-12 \text{ cm}}{4.0 \text{ cm}} = 3.0$$

The image is three times as large as the object and, because M is positive, upright.

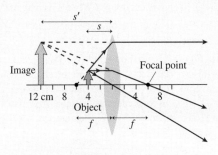

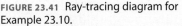

FIGURE 23.41 Ray-tracing diagram for Example 23.10.

Diverging Lenses

A lens thicker at the edges than in the middle is called a *diverging lens*. Figure 23.42 shows three important sets of rays passing through a diverging lens. These are based on Figures 23.33 and 23.34, where you saw that rays initially parallel to the axis diverge after passing through a diverging lens.

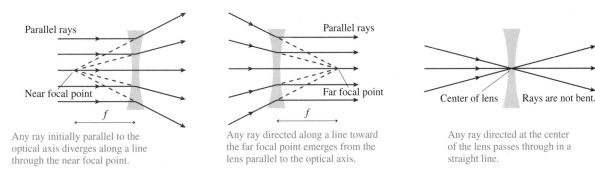

Any ray initially parallel to the optical axis diverges along a line through the near focal point.

Any ray directed along a line toward the far focal point emerges from the lens parallel to the optical axis.

Any ray directed at the center of the lens passes through in a straight line.

FIGURE 23.42 Three important sets of rays passing through a thin, diverging lens.

Ray tracing follows the steps of Tactics Box 23.2 for a converging lens *except* that two of the three special rays in step 4 are different.

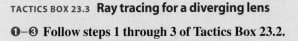

TACTICS BOX 23.3 Ray tracing for a diverging lens

❶–❸ **Follow steps 1 through 3 of Tactics Box 23.2.**
❹ **Draw the three "special rays" from the tip of the arrow.** Use a straightedge.

    a. A ray parallel to the axis diverges along a line through the near focal point.

    b. A ray along a line toward the far focal point emerges parallel to the axis.

    c. A ray through the center of the lens does not bend.

❺ **Trace the diverging rays backward.** The point from which they are diverging is the image point, which is always a virtual image.
❻ **Measure the image distance s'.** This will be a negative number.

EXAMPLE 23.11 Demagnifying a flower
A diverging lens with a focal length of 50 cm is placed 100 cm from a flower. Where is the image? What is its magnification?

MODEL The flower is in the object plane. Use ray tracing to locate the image.

VISUALIZE Figure 23.43 shows the ray-tracing diagram. The three special rays (labeled a, b, and c to match the Tactics Box) do not converge. However, they can be traced backward to an intersection ≈33 cm to the left of the lens. A virtual image is formed at $s' = -33$ cm with magnification

$$M = -\frac{s'}{s} = -\frac{-33 \text{ cm}}{100 \text{ cm}} = 0.33$$

The image, which can be seen by looking *through* the lens, is one-third the size of the object and upright.

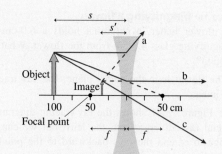

FIGURE 23.43 Ray-tracing diagram for Example 23.11.

ASSESS Ray tracing with a diverging lens is somewhat trickier than for a converging lens, so this example is worth careful study.

Diverging lenses *always* make virtual images and, for this reason, are rarely used alone. However, they have important applications when used in combination with other lenses. Cameras, eyepieces, and eyeglasses often incorporate diverging lenses.

Lens Combinations

Optical instruments, such as microscopes and cameras, are built with multiple lenses. There are many reasons for this having to do with image quality and the overall orientation and magnification of the image. A full analysis of optical instruments is beyond this text, but we can use the ideas of ray tracing to understand some of the basic ideas of lens combinations.

As an example, Figure 23.44 shows a telescope similar to the one with which Galileo discovered sunspots and the moons of Jupiter. It consists of a large converging lens, called the *objective,* and a smaller converging lens used as the *eyepiece.* The lenses are placed such that their focal points nearly coincide.

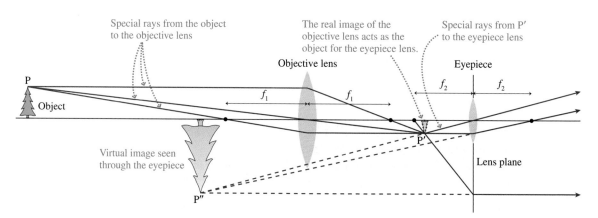

FIGURE 23.44 Ray-tracing diagram of a Galilean telescope.

The rays passing through the objective converge to a real image at P′, but they don't stop there. Instead, light rays *diverge* from P′ as they approach the second lens. As far as the eyepiece is concerned, the rays are coming from P′, and thus P′ is the object for the second lens. In other words, **the image of the first lens in a lens combination is the object for the second lens.**

The three special rays for the objective lens locate the image P′, but only one of these (parallel to the axis) is a special ray for the eyepiece. However, these aren't the only rays. Other rays will leave P′ at the correct angles to be the special rays for the eyepiece. That is, a new set of special rays is drawn from P′ to the second lens and used to find the final image point P″.

NOTE ▶ One ray seems to "miss" the eyepiece lens, but this isn't really a problem. First, we don't know the actual diameter of the lens. The lens diameter in the figure was an arbitrary choice, and the actual lens might be larger or smaller than shown. Second, the purpose of the special rays is to locate the point where *all* rays converge (or from which they diverge). Whether each special ray actually makes it through the lens isn't relevant. We can let the special rays refract as they cross the *lens plane,* regardless of whether the lens itself extends that far. ◀

The eyepiece acts as a magnifier because its object, point P′, is inside the focal point. Consequently, P″ is an enlarged, inverted, virtual image that is seen by looking through the eyepiece. The fact that a Galilean telescope produces an inverted image is not a problem in astronomy, but Galilean telescopes are not suitable for bird watching. Other telescope designs produce an upright image.

23.7 Thin Lenses: Refraction Theory

Ray tracing is a powerful visual approach for understanding image formation, but it doesn't provide precise information about the image location or image properties. We need to develop a quantitative relationship between the object distance s and the image distance s'. We will first analyze the refraction at a single spherical surface, using a method similar to our analysis of image formation by a plane refracting surface, in Section 23.4. Then we'll put two spherical surfaces together to form a lens.

To begin, Figure 23.45 shows a *spherical* boundary between two transparent media with indices of refraction n_1 and n_2. The sphere has radius of curvature R and is centered at point C. Consider a ray that leaves object point P at angle α and later, after refracting, reaches point P′. Figure 23.45 has exaggerated the angles to make the picture clear, but we will restrict our analysis to *paraxial rays* traveling nearly parallel to the axis. For paraxial rays, all the angles are small and we can use the small-angle approximation.

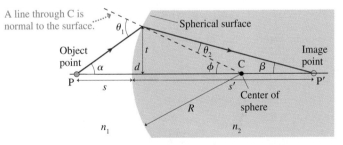

FIGURE 23.45 Image formation due to refraction at a spherical surface. The angles are exaggerated.

The ray from P is incident on the boundary at angle θ_1 and refracts into medium n_2 at angle θ_2, both measured from the normal to the surface at the point of incidence. Snell's law for the refraction is $n_1 \sin\theta_1 = n_2 \sin\theta_2$, which in the small-angle approximation is

$$n_1\theta_1 = n_2\theta_2 \tag{23.15}$$

You can see from the geometry of Figure 23.45 that angles α, β, and ϕ are related by

$$\theta_1 = \alpha + \phi$$
$$\theta_2 = \phi - \beta \tag{23.16}$$

Using these expressions in Equation 23.15, we can write Snell's law as

$$n_1(\alpha + \phi) = n_2(\phi - \beta) \tag{23.17}$$

This is one important relationship between the angles.

The line of height t, from the axis to the point of incidence, is the vertical leg of three different right triangles having vertices at points P, C, and P′. Consequently,

$$\tan\alpha \approx \alpha = \frac{t}{s + d} \qquad \tan\beta \approx \beta = \frac{t}{s' - d} \qquad \tan\phi \approx \phi = \frac{t}{R - d} \tag{23.18}$$

But $d \to 0$ for paraxial rays, thus

$$\alpha = \frac{t}{s} \qquad \beta = \frac{t}{s'} \qquad \phi = \frac{t}{R} \tag{23.19}$$

This is the second important relationship that comes from the geometry of Figure 23.45.

If we use the angles of Equation 23.19 in Equation 23.17, we find

$$n_1\left(\frac{t}{s} + \frac{t}{R}\right) = n_2\left(\frac{t}{R} - \frac{t}{s'}\right) \tag{23.20}$$

The t cancels, and we can rearrange Equation 23.20 to read

$$\frac{n_1}{s} + \frac{n_2}{s'} = \frac{n_2 - n_1}{R} \tag{23.21}$$

Equation 23.21 is independent of angle α. Consequently, **all paraxial rays that leave point P later converge at point P′.** If an object is located at distance s from a spherical refracting surface, an image will be formed at distance s' given by Equation 23.21.

Equation 23.21 was derived for a surface that is convex toward the object point, and the image is real. However, the result is also valid for virtual images or for surfaces that are concave toward the object point as long as we adopt the *sign convention* shown in Table 23.3.

TABLE 23.3 Sign convention for refracting surfaces

	Positive	Negative
R	Convex toward the object	Concave toward the object
s'	Real image, opposite side from object	Virtual image, same side as object

Section 23.4 considered image formation due to refraction by a plane surface. There we found (in Equation 23.13) an image distance $s' = (n_2/n_1)s$. A plane can be thought of as a sphere in the limit $R \to \infty$, so we should be able to reach the same conclusion from Equation 23.21. As $R \to \infty$, the term $(n_2 - n_1)/R \to 0$ and Equation 23.21 becomes $s' = -(n_2/n_1)s$. This seems to differ from Equation 23.13, but it doesn't really. Equation 23.13 gives the actual distance to the image. Equation 23.21 is based on a sign convention in which virtual images have negative image distances, hence the minus sign.

EXAMPLE 23.12 Image formation inside a glass rod
One end of a 4.0-cm-diameter glass rod is shaped as a hemisphere. A small light bulb is 6.0 cm from the end of the rod. Where is the bulb's image located?

MODEL Model the light bulb as a point source of light and consider the paraxial rays that refract into the glass rod.

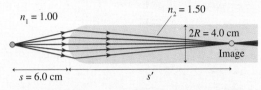

FIGURE 23.46 The curved surface refracts the light to form a real image.

VISUALIZE Figure 23.46 shows the situation. $n_1 = 1.00$ for air and $n_2 = 1.50$ for glass.

SOLVE The radius of the surface is half the rod diameter, so $R = 2.0$ cm. Equation 23.21 is

$$\frac{1.00}{6.0 \text{ cm}} + \frac{1.50}{s'} = \frac{1.50 - 1.00}{2.0 \text{ cm}} = \frac{0.50}{2.0 \text{ cm}}$$

Solving for the image distance s' gives

$$\frac{1.50}{s'} = \frac{0.50}{2.0 \text{ cm}} - \frac{1.00}{6.0 \text{ cm}} = 0.0833 \text{ cm}^{-1}$$

$$s' = \frac{1.50}{0.0833} = 18 \text{ cm}$$

ASSESS This is a real image located 18 cm inside the glass rod.

EXAMPLE 23.13 A goldfish in a bowl
A goldfish lives in a spherical fish bowl 50 cm in diameter. If the fish is 10 cm from the near edge of the bowl, where does the fish appear when viewed from the outside?

MODEL Model the fish as a point source and consider the paraxial rays that refract from the water into the air. The thin glass wall has little effect and will be ignored.

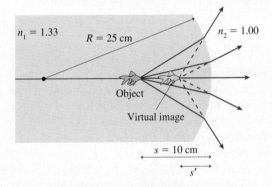

FIGURE 23.47 The curved surface of a fish bowl produces a virtual image of the fish.

VISUALIZE Figure 23.47 shows the rays refracting *away* from the normal as they move from the water into the air. We expect to find a virtual image at a distance less than 10 cm.

SOLVE The object is in the water, so $n_1 = 1.33$ and $n_2 = 1.00$. The inner surface is concave (you can remember "concave" because it's like looking into a cave), so $R = -25$ cm. The object distance is $s = 10$ cm. Thus Equation 23.21 is

$$\frac{1.33}{10 \text{ cm}} + \frac{1.00}{s'} = \frac{1.00 - 1.33}{-25 \text{ cm}} = \frac{0.33}{25 \text{ cm}}$$

Solving for the image distance s' gives

$$\frac{1.00}{s'} = \frac{0.33}{25} - \frac{1.33}{10} = -0.119 \text{ cm}^{-1}$$

$$s' = \frac{1.00}{-0.119} = -8.35 \text{ cm}$$

ASSESS The image is virtual, located to the left of the boundary. A person looking into the bowl will see a fish that appears to be 8.35 cm from the edge of the bowl.

STOP TO THINK 23.5 Which of these actions will move the image point P′ farther from the boundary? More than one may work.

a. Increase the radius of curvature R.
b. Increase the index of refraction n.
c. Increase the object distance s.
d. Decrease the radius of curvature R.
e. Decrease the index of refraction n.
f. Decrease the object distance s.

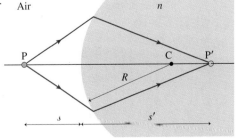

Lenses

15.10–15.12   Activ ONLINE Physics

A lens consists of *two* spherical surfaces having radii of curvature R_1 and R_2 and thickness t. The lens material has index of refraction n, and for simplicity we'll assume that the lens is surrounded by air. We'll analyze the converging lens shown in Figure 23.48, but our results will apply to any lens if we use the sign convention given above in Table 23.3.

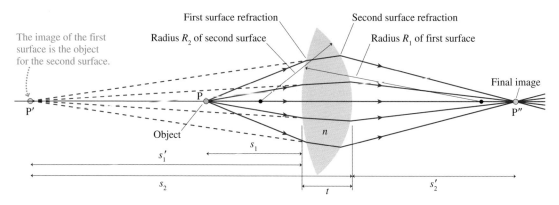

FIGURE 23.48 Image formation by a lens.

The object at point P is distance s_1 to the left of the lens. The first surface of the lens, of radius R_1, refracts the rays from P to create an image at point P'. We can use Equation 23.21 for a spherical surface to find the image distance s_1':

$$\frac{1}{s_1} + \frac{n}{s_1'} = \frac{n-1}{R_1} \tag{23.22}$$

where we used $n_1 = 1$ for the air and $n_2 = n$ for the lens. We'll assume that the image P' is a virtual image, but this assumption isn't essential to the outcome.

The image P' of the first surface becomes the object for the second surface. Object distance s_2 from P' to the second surface looks like it should be $s_2 = s_1' + t$, but P' is a virtual image of the first surface, so s_1' is a *negative* number. Thus the distance to the second surface is $s_2 = |s_1'| + t = t - s_1'$. We can find the image of P' by a second application of Equation 23.21, but with a switch. The rays are incident on the surface from within the lens, so this time $n_1 = n$ and $n_2 = 1$. Consequently,

$$\frac{n}{t - s_1'} + \frac{1}{s_2'} = \frac{1-n}{R_2} \tag{23.23}$$

For a *thick lens*, where the thickness t is not negligible, we can solve Equations 23.22 and 23.23 in sequence to find the position of the image point P''. But our primary interest is the *thin lens*. In the limit $t \rightarrow 0$, Equation 23.23 becomes

$$-\frac{n}{s_1'} + \frac{1}{s_2'} = \frac{1-n}{R_2} = -\frac{n-1}{R_2} \tag{23.24}$$

Our goal is to find the distance s_2' to point P'', the image produced by the lens as a whole. This goal is easily reached if we simply add Equations 23.22 and 23.24, eliminating s_1' and giving

$$\frac{1}{s_1} + \frac{1}{s_2'} = \frac{n-1}{R_1} - \frac{n-1}{R_2} = (n-1)\left(\frac{1}{R_1} - \frac{1}{R_2}\right) \tag{23.25}$$

The numerical subscripts on s_1 and s_2' no longer serve a purpose. If we replace s_1 by s, the object distance of the lens, and s_2' by s', the image distance, Equation 23.25 becomes the *thin-lens equation*

$$\frac{1}{s} + \frac{1}{s'} = \frac{1}{f} \quad \text{(thin-lens equation)} \qquad (23.26)$$

where the *focal length* of the lens is given by

$$\frac{1}{f} = (n - 1)\left(\frac{1}{R_1} - \frac{1}{R_2}\right) \quad \text{(lens maker's equation)} \qquad (23.27)$$

Equation 23.27 is known as the *lens maker's equation.* It allows you to determine the focal length from the shape of a lens and the material used to make it.

We can verify that this expression for f really is the focal length of the lens by recalling that rays initially parallel to the optical axis pass through the focal point on the far side. In fact, this was our *definition* of the focal length of a lens. Parallel rays must come from an object extremely far away, with object distance $s \to \infty$. In that case, $1/s = 0$ and Equation 23.26 tells us that the parallel rays will converge at distance $s' = f$ on the far side of the lens, exactly as expected.

Similarly, Equation 23.26 gives $1/s' = 0$, or $s' \to \infty$, for a point source of light at object distance $s = f$. In other words, an object at the near focal point produces an image infinitely far away, so the rays leave the lens traveling parallel to the axis. This is what we saw in Figure 23.35b. Thus the quantity f in Equation 23.27 really does represent the focal length of the lens.

We derived the thin-lens equation and the lens maker's equation from the specific lens geometry shown in Figure 23.48, but the results are valid for any lens as long as all quantities are given appropriate signs. The sign convention used with Equations 23.26 and 23.27 is given in Table 23.4.

TABLE 23.4 Sign convention for thin lenses

	Positive	Negative
R_1, R_2	Convex toward the object	Concave toward the object
f	Converging lens, thicker in center	Diverging lens, thinner in center
s'	Real image, opposite side from object	Virtual image, same side as object

EXAMPLE 23.14 Focal length of a meniscus lens
What is the focal length of the glass *meniscus lens* shown in Figure 23.49? Is this a converging or diverging lens?

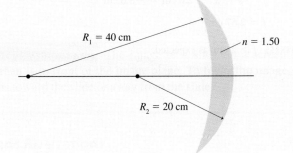

$R_1 = 40$ cm

$n = 1.50$

$R_2 = 20$ cm

FIGURE 23.49 A meniscus lens.

SOLVE If the object is on the left, then the first surface has $R_1 = -40$ cm (concave toward the object) and the second surface has $R_2 = -20$ cm (also concave toward the object). The index of refraction of glass is $n = 1.50$, so the lens maker's equation is

$$\frac{1}{f} = (n - 1)\left(\frac{1}{R_1} - \frac{1}{R_2}\right) = (1.50 - 1)\left(\frac{1}{-40\text{ cm}} - \frac{1}{-20\text{ cm}}\right)$$

$$= 0.0125 \text{ cm}^{-1}$$

Inverting this expression gives $f = 80$ cm. This is a converging lens, as seen both from the positive value of f and from the fact that the lens is thicker in the center.

(a) Chromatic aberration

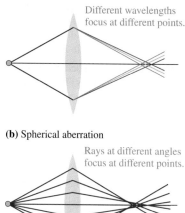

Different wavelengths focus at different points.

(b) Spherical aberration

Rays at different angles focus at different points.

FIGURE 23.53 Chromatic aberration and spherical aberration prevent simple lenses from forming perfect images.

First, any lens has dispersion. That is, its index of refraction varies slightly with wavelength. Because the index of refraction for violet light is larger than for red light, a lens's focal length is shorter for violet light than for red light. Consequently, different colors of light come to a focus at slightly different distances from the lens. If red light is sharply focused on a viewing screen, then blue and violet wavelengths are not well focused. This imaging error, illustrated in Figure 23.53a, is called **chromatic aberration.**

Second, our analysis of image formation was based on paraxial rays traveling nearly parallel to the optical axis. This assumption allowed us to use the small-angle approximation. A more exact analysis, taking all the rays into account, finds that rays incident on the outer edges of a spherical surface are not focused at exactly the same point as rays incident near the center. This imaging error, shown in Figure 23.53b, is called **spherical aberration.** Spherical aberration, which causes the image to be slightly blurred, gets worse as the lens diameter increases.

Fortunately, the aberrations of a converging lens and a diverging lens are in opposite directions. When a converging lens and a diverging lens are used in combination, their aberrations tend to cancel. Thus a combination lens, such as the one in Example 23.17, can produce a much sharper focus than can a single lens with the equivalent focal length. Consequently, cameras, microscopes, and other optical equipment use combination lenses rather than single lenses.

23.8 The Resolution of Optical Instruments

16.8 Actïv Physics

According to the ray model of light, a perfect lens (one with no aberrations) should be able to form a perfect image. But the ray model of light, while a very good model for lenses, is not an absolutely correct description of light. If we look closely, the wave aspects of light haven't entirely disappeared. In fact, the performance of optical equipment is limited by the waviness of light.

Figure 23.54a shows a plane wave being focused by a lens of diameter D. Only those waves passing *through* the lens can be focused, so the lens acts like a circular aperture in an opaque barrier. In other words, the lens both *focuses and diffracts* light waves. Figure 23.54b separates these two effects by modeling a real lens as an "ideal" diffractionless lens behind a circular aperture of diameter D.

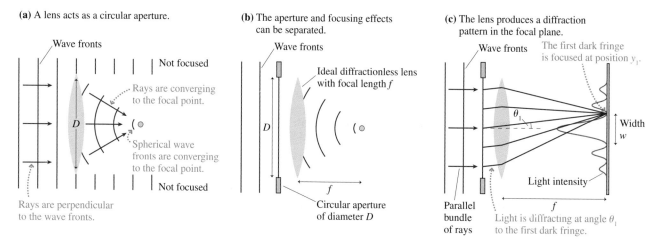

(a) A lens acts as a circular aperture.

Wave fronts

Not focused

Rays are converging to the focal point.

D

Spherical wave fronts are converging to the focal point.

Not focused

Rays are perpendicular to the wave fronts.

(b) The aperture and focusing effects can be separated.

Wave fronts

Ideal diffractionless lens with focal length f

D

f

Circular aperture of diameter D

(c) The lens produces a diffraction pattern in the focal plane.

Wave fronts

The first dark fringe is focused at position y_1.

θ_1

Width w

Light intensity

Parallel bundle of rays

f

Light is diffracting at angle θ_1 to the first dark fringe.

FIGURE 23.54 A lens both focuses and diffracts the light passing through.

You learned in Chapter 22 that a circular aperture produces a diffraction pattern with a bright central maximum surrounded by dimmer circular fringes. A converging lens brings parallel light rays to a focus at distance f. Consequently, as

Figure 23.54c shows, a lens behind a circular aperture collects all the light rays diffracting at angle θ and brings these rays together in the focal plane of the lens. The net result is that the image of a parallel bundle of rays is not a perfect point but, instead, a circular diffraction pattern.

The angle to the first minimum of a circular diffraction pattern is $\theta_1 = 1.22\lambda/D$. The ray that passes through the center of a lens is not bent, so Figure 23.54c uses this ray to show that the position of the dark fringe is $y_1 = f\tan\theta_1 \approx f\theta_1$. Thus the width of the central maximum in the focal plane is

$$w = 2f\theta_1 = \frac{2.44\lambda f}{D} \quad \text{(minimum spot size)} \qquad (23.28)$$

This is the **minimum spot size** to which a lens can focus light.

Lenses are often limited by aberrations, so not all lenses can focus light to a spot this small. A well-crafted lens, for which this is the minimum spot size, is called a *diffraction-limited lens*. No optical design can overcome the spreading of light due to diffraction, and it is because of this spreading that the image point has a minimum spot size.

For various reasons, it is difficult to produce a diffraction-limited lens having a focal length less than its diameter. That is, $f \geq D$ for any realistic lens. This implies that **the smallest diameter to which you can focus a spot of light, no matter how hard you try, is $w_{min} \approx 2.5\lambda$.** This is a fundamental limit on the performance of optical equipment. Diffraction has very real consequences!

One example of these consequences is found in the manufacturing of integrated circuits. Integrated circuits are made by creating a "mask" that shows all the components and their connections. A lens images this mask onto the surface of a semiconductor wafer that has been coated with a substance called *photoresist*. Bright areas in the mask expose the photoresist, and subsequent processing steps chemically etch away the exposed areas while leaving behind areas that had been in the shadows of the mask. This process is called photolithography.

The power of a microprocessor and the amount of memory in a memory chip depend on how small the circuit elements can be made. Diffraction dictates that a circuit element can be no smaller than the smallest spot to which light can be focused. That is, no feature on the chip can be smaller than roughly 2.5λ. If the mask is projected with ultraviolet light having $\lambda \approx 200 \text{ nm} = 0.2 \ \mu\text{m}$, then the smallest elements on a chip are about 0.50 μm wide. This is, in fact, just about the current limit of technology.

The size of the features in an integrated circuit is limited by the diffraction of light.

EXAMPLE 23.18 Looking at the stars

A 12-cm-diameter telescope lens has a focal length of 1.0 m. What is the diameter of the image of a star in the focal plane if the lens is diffraction limited *and* if the earth's atmosphere is not a limitation?

MODEL Stars are so far away that they appear as points in space. An ideal diffractionless lens would focus their light to an arbitrarily small point. Diffraction prevents this. Model the telescope lens as a 12-cm-diameter aperture in front of an ideal lens with a 1.0 m focal length.

SOLVE The minimum spot size in the focal plane of this lens is

$$w = \frac{2.44\lambda f}{D}$$

where D is the lens diameter. What is λ? Because stars emit white light, the *longest* wavelengths spread the most and determine the size of the image that is seen. If we use $\lambda = 700$ nm as the approximate upper limit of visible wavelengths, we find $w = 1.4 \times 10^{-5} \text{ m} = 14 \ \mu\text{m}$.

ASSESS This is certainly small, and it would appear as a point to your unaided eye. Nonetheless, the spot size would be easily noticed if it were recorded on film and enlarged. Turbulence and temperature effects in the atmosphere, the causes of the "twinkling" of stars, generally prevent ground-based telescopes from being this good, but space-based telescopes really are diffraction limited.

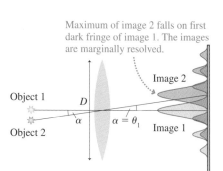

Maximum of image 2 falls on first dark fringe of image 1. The images are marginally resolved.

FIGURE 23.55 Two images that are marginally resolved.

Resolution

Suppose you point a telescope at two nearby stars in a galaxy far, far away. If you use the best possible detector, will you be able to distinguish separate images for the two stars, or will they blur into a single blob of light? A similar question could be asked of a microscope. Can two microscopic objects, very close together, be distinguished if sufficient magnification is used? Or is there some size limit at which they will blur together and never be separated? These are important questions about the **resolution** of optical instruments.

Because of diffraction, the image of a distant star is not a point but a circular diffraction pattern. Our question, then, really amounts to asking how close together two diffraction patterns can be before you can no longer distinguish them. One of the major scientists of the 19th century, Lord Rayleigh, studied this problem and suggested a reasonable rule that today is called **Rayleigh's criterion.**

Figure 23.55 shows two objects being imaged by a lens of diameter D. The angular separation between the objects, as seen from the lens, is α. Rayleigh's criterion states that

- The two objects are resolvable if $\alpha > \theta_1$, where $\theta_1 = 1.22\lambda/D$ is the angle of the first dark fringe in the circular diffraction pattern.
- The two objects are not resolvable if $\alpha < \theta_1$ because their diffraction patterns are too overlapped.
- The two objects are marginally resolvable if $\alpha = \theta_1$. The central maximum of one image falls exactly on top of the first dark fringe of the other image. This is the situation shown in the figure.

Figure 23.56 shows enlarged photographs of the images of two point sources. The images are circular diffraction patterns, not points. The two images are close but distinct where the objects are separated by $\alpha > \theta_1$. Two objects really were recorded in the photo on the right, but their separation is $\alpha < \theta_1$ and their images have blended together. In the middle photo, with $\alpha = \theta_1$, you can see that the two images are just barely resolved.

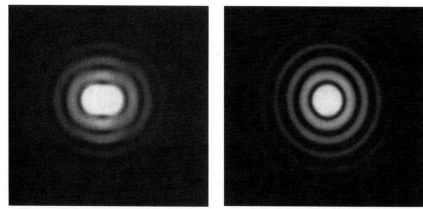

$\alpha > \theta_1$
Resolved

$\alpha = \theta_1$
Marginally resolved

$\alpha < \theta_1$
Not resolved

FIGURE 23.56 Enlarged photographs of the images of two closely spaced objects.

For telescopes, the angle $\theta_1 = 1.22\lambda/D$ is called the *angular resolution* of the telescope. The angular resolution is a function of the lens diameter and the wavelength; the magnification is not a factor. Two overlapped, unresolved images will remain overlapped and unresolved no matter what the magnification. For visible light, where λ is pretty much fixed, the only parameter over which the astronomer has any control is the diameter of the lens or mirror of the telescope. The urge to build ever larger telescopes is motivated, in part, by a desire to improve the angular resolution. (Another motivation is to increase the light-gathering power so as to see objects farther away.)

A microscope is rather like a telescope in reverse. The object is located at distance $s \approx f$ in front of the lens and the image is formed much farther away behind the lens. (Think how close the objective of a microscope is placed to the object and how far away, by comparison, the eyepiece is.) A microscope with an objective lens of diameter D can marginally resolve two objects separated by angle $\alpha = \theta_1 = 1.22\lambda/D$. At distance f, the physical separation d between the two objects is $s = f\tan\alpha \approx f\alpha$, where we've used the small-angle approximation. Thus the smallest separation of two objects that can be resolved by a microscope is

$$d_{min} = f\theta_1 = \frac{1.22\lambda f}{D} \approx \lambda \qquad (23.29)$$

In the last step we have used the fact, previously noted, that for any real lens the minimum value of f/D is ≈ 1.

The ultimate performance of a microscope is limited by the diffraction of light through the objective lens. Objects smaller than about one wavelength of light, roughly 1 μm, *cannot* be resolved by an optical microscope. This is a conclusion of fundamental importance. Because atoms are approximately 0.1 nm in diameter, vastly smaller than the wavelength of visible or even ultraviolet light, there is no hope of ever seeing atoms with an optical microscope. This limitation is not simply a matter of needing a better design or more precise components. It is a fundamental limit set by the wave nature of the light with which we see.

STOP TO THINK 23.7   Four diffraction-limited lenses focus plane waves of light with the same wavelength λ. Rank in order, from largest to smallest, the spot sizes w_1 to w_4.

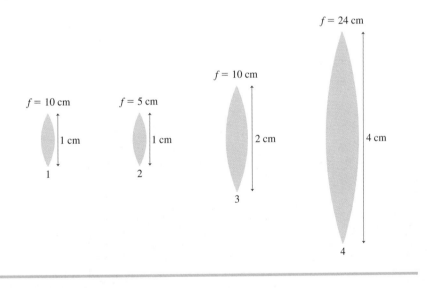

SUMMARY

The goal of Chapter 23 has been to understand and apply the ray model of light.

GENERAL PRINCIPLES

Reflection

Law of reflection: $\theta_r = \theta_i$

Reflection can be **specular** (mirror-like) or **diffuse** (from rough surfaces).

Plane mirrors: A virtual image is formed at P' with $s' = s$.

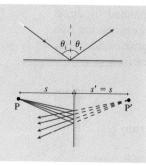

Refraction

Snell's law of refraction:
$n_1 \sin \theta_1 = n_2 \sin \theta_2$

Index of refraction is $n = c/v$. The ray is closer to the normal on the side with the larger index of refraction.

If $n_2 < n_1$, total internal reflection (TIR) occurs when the angle of incidence $\theta_1 > \theta_c = \sin^{-1}(n_2/n_1)$.

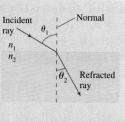

IMPORTANT CONCEPTS

The ray model of light

Light travels along straight lines, called **light rays,** at speed $v = c/n$.

A light ray continues forever unless an interaction with matter causes it to reflect, refract, scatter, or be absorbed.

Light rays come from **objects.** Each point on the object sends rays in all directions.

The eye sees an object (or an image) when diverging rays are collected by the pupil and focused on the retina.

▶ Ray optics is valid when lenses, mirrors, and apertures are larger than ≈1 mm in size.

Image formation

If rays diverge from P and interact with a lens or mirror so that the refracted/reflected rays *diverge* from P' and appear to come from P', then P' is a virtual image of P.

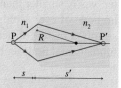

If rays diverge from P and interact with a lens so that the refracted rays *converge* at P', then P' is a real image of P.

Spherical surface: Object and image distances are related by

$$\frac{n_1}{s} + \frac{n_2}{s'} = \frac{n_2 - n_1}{R}$$

Plane surface: $R \to \infty$, so $|s'/s| = n_2/n_1$

APPLICATIONS

Ray tracing

3 special rays in 3 basic situations:

Converging lens   Converging lens   Diverging lens
Real image   Virtual image   Virtual image

Magnification $\quad M = -\dfrac{h'}{h} = -\dfrac{s'}{s}$

M is + for an upright image, − for inverted.

The height ratio is $h'/h = |M|$.

Thin lenses

The image and object distance are related by

$$\frac{1}{s} + \frac{1}{s'} = \frac{1}{f}$$

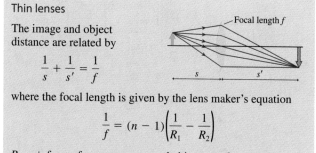

where the focal length is given by the lens maker's equation

$$\frac{1}{f} = (n - 1)\left(\frac{1}{R_1} - \frac{1}{R_2}\right)$$

R + for surface convex toward object   − for concave
f + for a converging lens   − for diverging
s' + for a real image   − for virtual

Lens combinations: Image of first is object for second.

Resolution of optical instruments

Because of diffraction, the **minimum spot size** of a lens of diameter D is $w = 2.44\lambda f/D$.

Rayleigh's criterion: Two objects separated by angle α are marginally resolvable if $\alpha = \theta_1 = 1.22\lambda/D$.

TERMS AND NOTATION

light ray	law of reflection	optical axis	lens plane
object	diffuse reflection	paraxial rays	real image
point source	virtual image	dispersion	object plane
parallel bundle	refraction	Rayleigh scattering	image plane
ray diagram	angle of refraction	lens	inverted image
camera obscura	Snell's law	ray tracing	upright image
aperture	total internal reflection	converging lens	chromatic aberration
magnification, m or M	(TIR)	focal point	spherical aberration
specular reflection	critical angle, θ_c	focal length, f	minimum spot size
angle of incidence	object distance, s	diverging lens	resolution
angle of reflection	image distance, s'	thin lens	Rayleigh's criterion

EXERCISES AND PROBLEMS

Exercises

Section 23.1 The Ray Model of Light

1. a. How long (in ns) does it take light to travel 1.0 m in vacuum?
   b. What distance does light travel in water, glass, and zircon during the time that it travels 1.0 m in vacuum?
2. A 5.0-cm-thick layer of oil is sandwiched between a 1.0-cm-thick sheet of glass and a 2.0-cm-thick sheet of polystyrene plastic. How long (in ns) does it take light incident perpendicular to the glass to pass through this 8.0-cm-thick sandwich?
3. A point source of light illuminates an aperture 2.0 m away. A 12.0-cm-wide bright patch of light appears on a screen 1.0 m behind the aperture. How wide is the aperture?
4. Figure Ex23.4 is the top view of a room. Red and green light bulbs separated by 0.25 m shine through the door and illuminate the back wall. Over what range of x is the back wall illuminated by (a) the red and (b) the green light?

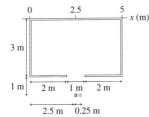

FIGURE EX23.4

5. A student has built a 20-cm-long pinhole camera for a science fair project. She wants to photograph the Washington Monument, which is 167 m (550 ft) tall, and to have the image on the film be 5.0 cm high. How far should she stand from the Washington Monument?

Section 23.2 Reflection

6. The mirror in Figure Ex23.6 deflects a horizontal laser beam by 60°. What is the angle ϕ?

FIGURE EX23.6

7. A light ray leaves point A in Figure Ex23.7, reflects from the mirror, and reaches point B. How far below the top edge does the ray strike the mirror?

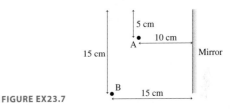

FIGURE EX23.7

8. You need to place a mirror on the left wall of Figure Ex23.8 so that the reflected light from the bulb exactly fills the right wall.
   a. What is the proper height of the mirror, and how far below the ceiling should the top edge be?
   b. A 10-cm-tall screen to the right of the bulb prevents the bulb from illuminating the right wall directly. What is the height of the screen's shadow on the right wall? Consider only the shadow due to light coming directly from the bulb, not light reflected by the mirror.

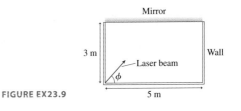

FIGURE EX23.8

9. At what angle ϕ should the laser beam in Figure Ex23.9 be aimed at the mirrored ceiling in order to hit the midpoint of the far wall?

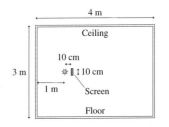

FIGURE EX23.9

10. Figure Ex23.10 is the top view of a room. As you walk along the wall opposite the mirror (i.e., along the *x*-axis), over what range of *x* can you see the entire green arrow in the mirror?

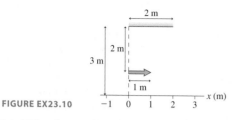

FIGURE EX23.10

11. It is 165 cm from your eyes to your toes. You're standing 200 cm in front of a tall mirror. How far is it from your eyes to the image of your toes?

Section 23.3 Refraction

12. A 1.0-cm-thick layer of water stands on a horizontal slab of glass. A light ray in the air is incident on the water 60° from the normal. What is the ray's direction of travel in the glass?

13. A diamond is underwater. A light ray enters one face of the diamond, then travels at an angle of 30° with respect to the normal. What was the ray's angle of incidence on the diamond?

14. An underwater diver sees the sun 50° above horizontal. How high is the sun above the horizon to a fisherman in a boat above the diver?

15. A laser beam in air is incident on a liquid at an angle of 37° with respect to the normal. The laser beam's angle in the liquid is 26°. What is the liquid's index of refraction?

16. The glass core of an optical fiber has an index of refraction 1.60. The index of refraction of the cladding is 1.48. What is the maximum angle a light ray can make with the wall of the core if it is to remain inside the fiber?

17. A thin glass rod is submerged in oil. What is the critical angle for light traveling inside the rod?

Section 23.4 Image Formation by Refraction

18. A fish in a flat-sided aquarium sees a can of fish food on the counter. To the fish's eye, the can looks to be 30 cm outside the aquarium. What is the actual distance between the can and the aquarium? (You can ignore the thin glass wall of the aquarium.)

19. A biologist keeps a specimen of his favorite beetle embedded in a cube of polystyrene plastic. The hapless bug appears to be 2.0 cm within the plastic. What is the beetle's actual distance beneath the surface?

20. A 150-cm-tall diver is standing completely submerged on the bottom of a swimming pool full of water. You are sitting on the end of the diving board, almost directly over her. How tall does the diver appear to be?

21. To a fish in an aquarium, the 4.00-mm-thick walls appear to be only 3.50 mm thick. What is the index of refraction of the walls?

Section 23.5 Color and Dispersion

22. A sheet of glass has $n_{red} = 1.52$ and $n_{violet} = 1.55$. A narrow beam of white light is incident on the glass at 30°. What is the angular spread of the light inside the glass?

23. A hydrogen discharge lamp emits light with two prominent wavelengths: 656 nm (red) and 486 nm (blue). The light enters a flint-glass prism perpendicular to one face and then refracts through the hypotenuse back into the air. The angle between these two faces is 35°.
    a. Use Figure 23.29 to estimate to ±0.002 the index of refraction of flint glass at these two wavelengths.
    b. What is the angle (in degrees) between the red and blue light as it leaves the prism?

24. A narrow beam of white light is incident on a sheet of quartz. The beam disperses in the quartz, with red light ($\lambda \approx 700$ nm) traveling at an angle of 26.3° with respect to the normal and violet light ($\lambda \approx 400$ nm) traveling at 25.7°. The index of refraction of quartz for red light is 1.45. What is the index of refraction of quartz for violet light?

25. Infrared telescopes, which use special infrared detectors, are able to peer farther into star-forming regions of the galaxy because infrared light is not scattered as strongly as is visible light by the tenuous clouds of hydrogen gas from which new stars are created. For what wavelength of light is the scattering only 1% that of light with a visible wavelength of 500 nm?

Section 23.6 Thin Lenses: Ray Tracing

26. An object is 20 cm in front of a converging lens with a focal length of 10 cm. Use ray tracing to determine the location of the image. Is the image upright or inverted?

27. An object is 30 cm in front of a converging lens with a focal length of 10 cm. Use ray tracing to determine the location of the image. Is the image upright or inverted?

28. An object is 6 cm in front of a converging lens with a focal length of 10 cm. Use ray tracing to determine the location of the image. Is the image upright or inverted?

29. An object is 15 cm in front of a diverging lens with a focal length of −10 cm. Use ray tracing to determine the location of the image. Is the image upright or inverted?

Section 23.7 Thin Lenses: Refraction Theory

30. Find the focal length of the glass lens in Figure Ex23.30.

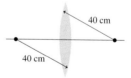

FIGURE EX23.30

31. Find the focal length of the planoconvex polystyrene plastic lens in Figure Ex23.31.

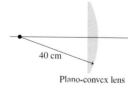

FIGURE EX23.31   Plano-convex lens

32. Find the focal length of the glass lens in Figure Ex23.32.

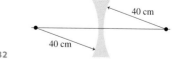

FIGURE EX23.32

33. Find the focal length of the meniscus polystyrene plastic lens in Figure Ex23.33.

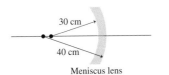

FIGURE EX23.33 Meniscus lens

34. A goldfish lives in a 50-cm-diameter spherical fish bowl. The fish sees a cat watching it. If the cat's face is 20 cm from the edge of the bowl, how far from the edge does the fish see it as being? (You can ignore the thin glass wall of the bowl.)

35. An air bubble inside an 8.0-cm-diameter plastic ball is 2.0 cm from the surface. As you look at the ball with the bubble turned toward you, how far beneath the surface does the bubble appear to be?

Section 23.8 The Resolution of Optical Instruments

36. A scientist needs to focus a helium-neon laser beam ($\lambda = 633$ nm) to a 10-μm-diameter spot 8.0 cm behind the lens. What focal-length lens should she use? What minimum diameter must the lens have?

37. Two light bulbs are 1.0 m apart. From what distance can these light bulbs be marginally resolved by a small telescope with a 4.0-cm-diameter objective lens? Assume that the lens is diffraction limited and $\lambda = 600$ nm.

Problems

38. An advanced computer sends information to its various parts via infrared light pulses traveling through silicon fibers. To acquire data from memory, the central processing unit sends a light-pulse request to the memory unit. The memory unit processes the request, then sends a data pulse back to the central processing unit. The memory unit takes 0.5 ns to process a request. If the information has to be obtained from memory in 2.0 ns, what is the maximum distance the memory unit can be from the central processing unit?

39. A red ball is placed at point A in Figure P23.39.
    a. How many images are seen by an observer at point O?
    b. Where is each image located?
    c. Draw a ray diagram showing the formation of each image.

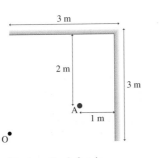

FIGURE P23.39

40. A laser beam is incident on the left mirror in Figure P23.40. Its initial direction is parallel to a line that bisects the mirrors. What is the angle ϕ of the reflected laser beam?

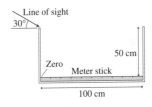

FIGURE P23.40

41. The place you get your hair cut has two nearly parallel mirrors 5.0 m apart. As you sit in the chair, your head is 2.0 m from the nearer mirror. Looking toward this mirror, you first see your face and then, farther away, the back of your head. (The mirrors need to be slightly nonparallel for you to be able to see the back of your head, but you can treat them as parallel in this problem.) How far away does the back of your head appear to be? Neglect the thickness of your head.

42. You're helping with an experiment in which a vertical cylinder will rotate about its axis by a very small angle. You need to devise a way to measure this angle. You decide to use what is called an *optical lever*. You begin by mounting a small mirror on top of the cylinder. A laser 5.0 m away shoots a laser beam at the mirror. Before the experiment starts, the mirror is adjusted to reflect the laser beam directly back to the laser. Later, you measure that the reflected laser beam, when it returns to the laser, has been deflected sideways by 2.0 mm. How many degrees has the cylinder rotated?

43. A 1.0-cm-thick layer of water stands on a horizontal slab of glass. Light from a source within the glass is incident on the glass-water boundary. What is the maximum angle of incidence for which the light ray can emerge into the air above the water?

44. A microscope is focused on a black dot. When a 1.00-cm-thick piece of plastic is placed over the dot, the microscope objective has to be raised 0.40 cm to bring the dot back into focus. What is the index of refraction of the plastic?

45. What is the angle of incidence in air of a light ray whose angle of refraction in glass is half the angle of incidence?

46. A meter stick lies on the bottom of a 100-cm-long tank with its zero mark against the left edge. You look into the tank at a 30° angle, with your line of sight just grazing the upper left edge of the tank. What mark do you see on the meter stick if the tank is (a) empty, (b) half full of water, and (c) completely full of water?

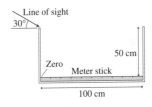

FIGURE P23.46

47. The 80-cm-tall, 65-cm-wide tank shown in Figure P23.47 is completely filled with water. The tank has marks every 10 cm along one wall, and the 0 cm mark is barely submerged. As you stand beside the opposite wall, your eye is level with the top of the water.
    a. Can you see the marks from the top of the tank (the 0 cm mark) going down, or from the bottom of the tank (the 80 cm mark) coming up? Explain.
    b. Which is the lowest or highest mark, depending on your answer to part a, that you can see?

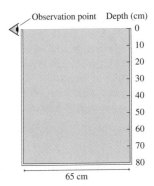

FIGURE P23.47

48. A 4.0-m-wide swimming pool is filled to the top. The bottom of the pool becomes completely shaded in the afternoon when the sun is 20° above the horizon. How deep is the pool?

49. It's nighttime, and you've dropped your goggles into a swimming pool that is 3.0 m deep. If you hold a laser pointer 1.0 m above the edge of the pool, you can illuminate the goggles if the laser beam enters the water 2.0 m from the edge. How far are the goggles from the edge of the pool?

50. Shown from above in Figure P23.50 is one corner of a rectangular box filled with water. A laser beam starts 10 cm from side A of the container and enters the water at position x. You can ignore the thin walls of the container.
    a. If $x = 15$ cm, does the laser beam refract back into the air through side B or reflect from side B back into the water? Determine the angle of refraction or reflection.
    b. Repeat part a for $x = 25$ cm.
    c. Find the minimum value of x for which the laser beam passes through side B and emerges into the air.

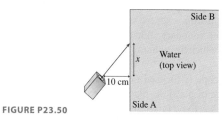

FIGURE P23.50

51. A fish is 20 m from the shore of a lake. A bonfire is burning on the edge of the lake nearest the fish.
    a. Does the fish need to be shallow (just below the surface) or very deep to see the light from the bonfire? Explain.
    b. What is the deepest or shallowest, depending on your answer to part a, that the fish can be and still see light from the fire?

52. One of the contests at the school carnival is to throw a spear at an underwater target lying flat on the bottom of a pool. The water is 1.0 m deep. You're standing on a small stool that places your eyes 3.0 m above the bottom of the pool. As you look at the target, your gaze is 30° below horizontal. At what angle below horizontal should you throw the spear in order to hit the target? Your raised arm brings the spear point to the level of your eyes as you throw it, and over this short distance you can assume that the spear travels in a straight line rather than a parabolic trajectory.

53. A narrow beam of white light is incident at 30° on a 10.0-cm-thick piece of glass. The rainbow of dispersed colors spans 1.00 mm on the bottom surface of the glass. The index of refraction for deep red light is 1.513. What is the index of refraction for deep violet light?

54. White light is incident onto a 30° prism at the 40° angle shown in Figure P23.54. Violet light emerges perpendicular to the rear face of the prism. The index of refraction of violet light in this glass is 2.0% larger than the index of refraction of red light. At what angle ϕ does red light emerge from the rear face?

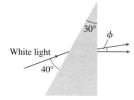

FIGURE P23.54

55. a. What is the smallest angle θ_1 for which a laser beam will undergo TIR on the hypotenuse of this glass prism?
    b. After reflecting from the hypotenuse at angle θ_c, the laser beam exits the prism through the bottom face. Does it exit to the right or to the left of the normal? At what angle?

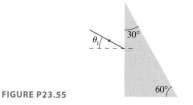

FIGURE P23.55

56. There's one angle of incidence β onto a prism for which the light inside an isosceles prism travels parallel to the base and emerges at angle β.
    a. Find an expression for β in terms of the prism's apex angle α and index of refraction n.
    b. A laboratory measurement finds that $\beta = 52.2°$ for a prism that is shaped as an equilateral triangle. What is the prism's index of refraction?

FIGURE 23.56

57. A 6.0-cm-diameter zircon sphere has an air bubble exactly in the center. As you look into the sphere, how far beneath the surface does the bubble appear to be?

58. Parallel light rays enter a transparent sphere along a line passing through the center of the sphere. The rays come to a focus on the far surface of the sphere. What is the sphere's index of refraction?

59. A 2.0-cm-tall object is 40 cm in front of a converging lens that has a 20 cm focal length.
    a. Use ray tracing to find the position and height of the image. To do this accurately use a ruler or paper with a grid. Determine the image distance and image height by making measurements on your diagram.
    b. Calculate the image position and height. Compare with your ray-tracing answers in part a.

60. A 1.0-cm-tall object is 10 cm in front of a converging lens that has a 30 cm focal length.
    a. Use ray tracing to find the position and height of the image. To do this accurately use a ruler or paper with a grid. Determine the image distance and image height by making measurements on your diagram.
    b. Calculate the image position and height. Compare with your ray-tracing answers in part a.

61. A 2.0-cm-tall object is 15 cm in front of a converging lens that has a 20 cm focal length.
    a. Use ray tracing to find the position and height of the image. To do this accurately use a ruler or paper with a grid. Determine the image distance and image height by making measurements on your diagram.
    b. Calculate the image position and height. Compare with your ray-tracing answers in part a.

62. A 1.0-cm-tall object is 75 cm in front of a converging lens that has a 30 cm focal length.
    a. Use ray tracing to find the position and height of the image. To do this accurately use a ruler or paper with a grid. Determine the image distance and image height by making measurements on your diagram.

b. Calculate the image position and height. Compare with your ray-tracing answers in part a.

63. A 2.0-cm-tall object is 15 cm in front of a diverging lens that has a −20 cm focal length.
    a. Use ray tracing to find the position and height of the image. To do this accurately use a ruler or paper with a grid. Determine the image distance and image height by making measurements on your diagram.
    b. Calculate the image position and height. Compare with your ray-tracing answers in part a.

64. A 1.0-cm-tall object is 60 cm in front of a diverging lens that has a −30 cm focal length.
    a. Use ray tracing to find the position and height of the image. To do this accurately use a ruler or paper with a grid. Determine the image distance and image height by making measurements on your diagram.
    b. Calculate the image position and height. Compare with your ray-tracing answers in part a.

65. A 2.0-cm-diameter spider is 2.0 m from a wall. Determine the focal length and position (measured from the wall) of a lens that will make a half-size image of the spider on the wall.

66. A 2.0-cm-tall candle flame is 2.0 m from a wall. You happen to have a lens with a focal length of 32 cm. How many places can you put the lens to form a well-focused image of the candle flame on the wall? For each location, what is the height and orientation of the image?

67. a. Estimate the diameter of your eyeball.
    b. Bring this page up to the closest distance at which the text is sharp—not the closest at which you can still read it, but the closest at which the letters remain sharp. If you wear glasses or contact lenses, leave them on. This distance is called the *near point* of your (possibly corrected) eye. Record it.
    c. Estimate the effective focal length of your eye. The effective focal length includes the focusing due to the lens, the curvature of the cornea, and any corrections you wear. Ignore the effects of the fluid in your eye.

68. A slide projector needs to create a 98-cm-high image of a 2.0-cm-tall slide. The screen is 300 cm from the slide.
    a. What focal length does the lens need? Assume that it is a thin lens.
    b. How far should you place the lens from the slide?

69. An object is 60 cm from a screen. What are the radii of a symmetric converging plastic lens (i.e., two equally curved surfaces) that will form an image on the screen twice the height of the object?

70. Two converging lenses with focal lengths of 40 cm and 20 cm are 10 cm apart. A 2.0-cm-tall object is 15 cm in front of the 40-cm-focal-length lens.
    a. Use ray tracing to find the position and height of the image. To do this accurately use a ruler or paper with a grid. Determine the image distance and image height by making measurements on your diagram.
    b. Calculate the image position and height. Compare with your ray-tracing answers in part a.

71. A converging lens with a focal length of 40 cm and a diverging lens with a focal length of −40 cm are 160 cm apart. A 2.0-cm-tall object is 60 cm in front of the converging lens.
    a. Use ray tracing to find the position and height of the image. To do this accurately use a ruler or paper with a grid. Determine the image distance and image height by making measurements on your diagram.

b. Calculate the image position and height. Compare with your ray-tracing answers in part a.

72. High-power lasers are used to cut and weld materials by focusing the laser beam to a very small spot. This is like using a magnifying lens to focus the sun's light to a small spot that can burn things. As an engineer, you have designed a laser cutting device in which the material to be cut is placed 5.0 cm behind the lens. You have selected a high-power laser with a wavelength of 1.06 μm. (This distance and wavelength are both very typical values.) Your calculations indicate that the laser must be focused to a 5.0-μm-diameter spot in order to have sufficient power to make the cut. What is the minimum diameter of the lens you must install?

73. Once dark adapted, the pupil of your eye is approximately 7 mm in diameter. The headlights of an oncoming car are 120 cm apart. If the lens of your eye is diffraction limited, at what distance are the two headlights marginally resolved? Assume a wavelength of 600 nm. (Your eye is not really good enough to resolve headlights at this distance, due both to aberrations in the lens and to the size of the receptors in your retina, but it comes reasonably close.)

74. Alpha Centauri, the nearest star to our solar system, is 4.3 light years away. Assume that Alpha Centauri has a planet with an advanced civilization. Professor Dhg, at the planet's Astronomical Institute, wants to build a telescope with which he can find out if there are planets orbiting the sun.
    a. What is the minimum diameter for an objective lens that will just barely resolve Jupiter and the sun? The radius of Jupiter's orbit is 780 million km. Assume $\lambda = 600$ nm.
    b. Building a telescope of the necessary size does not appear to be a major problem. What practical difficulties might prevent Professor Dhg's experiment from succeeding?

75. Optical disk storage uses a small infrared laser ($\lambda \approx 800$ nm) to read, via reflected light, "pits" that are burned into a plastic surface.
    a. What is the smallest spot size to which the laser beam can be focused?
    b. Assume the pits are located on a two-dimensional square grid with a spacing 25% larger than the laser spot size. (Spacing them any closer would risk reading errors.) Each pit records 1 bit of information, and it takes 8 bits to form 1 byte, the standard unit of data storage. An optical disk has a usable surface area with an inner diameter of 4 cm and an outer diameter of 11 cm. How many megabytes (MB) of data can be stored on an optical disk?

Challenge Problems

76. Figure CP23.76 on the next page shows a light ray that travels from point A to point B. The ray crosses the boundary at position x, making angles θ_1 and θ_2 in the two media. Suppose that you did *not* know Snell's law.
    a. Write an expression for the *time t* it takes the light ray to travel from A to B. Your expression should be in terms of the distances a, b, and w; the variable x; and the indices of refraction n_1 and n_2.
    b. The time depends on x. There's one value of x for which the light travels from A to B in the shortest possible time. We'll call it x_{min}. Write an expression (but don't try to solve it!) from which x_{min} could be found.

c. Now, by using the geometry of the figure, derive Snell's law from your answer to part b.

You've proven that Snell's law is equivalent to the statement that "light traveling between two points follows the path that requires the shortest time." This interesting way of thinking about refraction is called *Fermat's principle*.

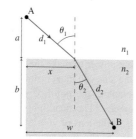

FIGURE CP23.76

77. A beam of white light enters a transparent material. Wavelengths for which the index of refraction is n are refracted at angle θ_2. Wavelengths for which the index of refraction is $n + \delta n$, where $\delta n \ll n$, are refracted at angle $\theta_2 + \delta\theta$.
    a. Show that the angular separation in radians is $\delta\theta = -\tan\theta_2(\delta n/n)$.
    b. A beam of white light is incident on a piece of glass at 30.0°. Deep violet light is refracted 0.28° more than deep red light. The index of refraction for deep red light is known to be 1.552. What is the index of refraction for deep violet light?

78. A fortune teller's "crystal ball" (actually just glass) is 10 cm in diameter. Her secret ring is placed 6 cm from the edge of the ball.
    a. An image of the ring appears on the opposite side of the crystal ball. How far is the image from the center of the ball?
    b. Draw a ray diagram showing the formation of the image.
    c. The crystal ball is removed and a thin lens is placed where the center of the ball had been. If the image is still in the same position, what is the focal length of the lens?

79. Consider a lens having index of refraction n_2 and surfaces with radii R_1 and R_2. The lens is immersed in a fluid that has index of refraction n_1.
    a. Derive a generalized lens maker's equation to replace Equation 23.27 when the lens is surrounded by a medium other than air. That is, when $n_1 \neq 1$.

b. A symmetric converging glass lens (i.e., two equally curved surfaces) has two surfaces with radii of 40 cm. Find the focal length of this lens in air and the focal length of this lens in water.

80. The closest you can bring an object to your eye and still see it clearly is called the eye's *near point*. A near point of 25 cm is considered normal vision. If your near point is more than 25 cm, you are *far sighted* and may need to have your vision corrected. Call your actual near point d. This is the shortest distance at which your unaided eye can focus. The purpose of a corrective lens is to create a virtual image at distance d of an object held at 25 cm. If you hold an object 25 cm away that you would like to focus on but can't, the lens creates a virtual image that you can see. Suppose your actual near point is 100 cm.
    a. What is the focal length of an eyeglass lens that will restore normal vision?
    b. Draw a ray diagram of your eye viewing an object 25 cm away.
    You can neglect the small distance between your eye and the lens.

81. The farthest distance at which you can see an object clearly is called the eye's *far point*. A far point infinitely far away is considered normal. (Stars, which are essentially at infinity, should appear as small, well-focused points of light.) If your far point is less than infinity, you are *near sighted* and may need to have your vision corrected. Call your actual far point d. This is the farthest distance at which your unaided eye can focus. The purpose of a corrective lens is to create a virtual image at distance d of an object that is very far away. If you look toward a very distant object that you would like to focus on but can't, the lens creates a virtual image that you can see. Suppose your actual far point is 200 cm.
    a. What is the focal length of an eyeglass lens that will restore normal vision?
    b. Draw a ray diagram of your eye viewing an object very far away.
    You can neglect the small distance between your eye and the lens.

82. A 1.0-cm-tall object is 110 cm from a screen. A diverging lens with focal length −20 cm is 20 cm in front of the object. What are the focal length and location of a second lens that will produce a well-focused, 2.0-cm-tall image on the screen?

STOP TO THINK ANSWERS

Stop to Think 23.1: c. The light spreads vertically as it goes through the vertical aperture. The light spreads horizontally due to different points on the horizontal light bulb.

Stop to Think 23.2: c. There's one image behind the vertical mirror and a second behind the horizontal mirror. A third image in the corner arises from rays that reflect twice, once off each mirror.

Stop to Think 23.3: a. The ray travels closer to the normal in both media 1 and 3 than in medium 2, so n_1 and n_3 are both larger than n_2. The angle is smaller in medium 3 than in medium 1, so $n_3 > n_1$.

Stop to Think 23.4: e. The rays from the object are diverging. Without a lens, the rays cannot converge to form any kind of image on the screen.

Stop to Think 23.5: a, e, or f. Any of these will increase the angle of refraction θ_2.

Stop to Think 23.6: Away from. You need to decrease s' to bring the image plane on the screen. s' is decreased by increasing s.

Stop to Think 23.7: $w_1 > w_4 > w_2 = w_3$. The spot size is proportional to f/D.

24 Modern Optics and Matter Waves

This image from a scanning tunneling microscope shows individual silicon atoms at the surface of a silicon crystal.

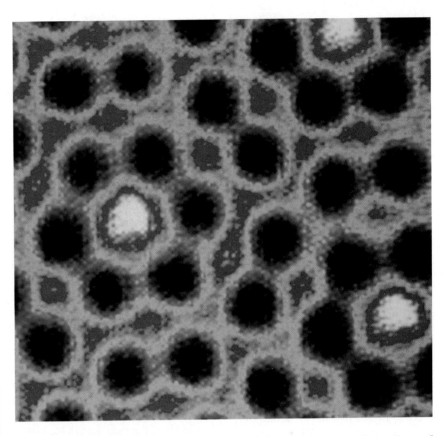

The scanning tunneling microscope is one of the most important inventions of the last 30 years. For the first time, we can "see" the structure of materials at the atomic level. The scanning tunneling microscope works by exploiting the wave properties of electrons.

Wave properties? Aren't electrons particles?

To answer this question we must look more closely at the fundamental properties of matter and light. Our journey through physics has brought us to about 1890, a little over a century ago. The physics of particles and waves was well understood by then, and it seemed that Newtonian physics would soon succeed in explaining all the phenomena of nature in terms of the particle and wave models. But trouble was on the horizon. Discoveries made during the last decade of the 19th century and the opening years of the 20th century refused to yield to a Newtonian analysis.

At the heart of the crisis was a breakdown of the basic particle and wave models. As physicists probed more deeply into the nature of light, they began to make observations that couldn't be reconciled with the wave model. Sometimes, as you will see, light refuses to act like a wave and seems more like a collection of particles. Even more troubling experiments found that electrons sometimes behave like waves.

These discoveries eventually led to a radical new theory of light and matter called *quantum physics*. We will return to study quantum physics more thoroughly

Some modern spectrometers are small enough to hold in your hand. (The rainbow has been painted onto this photograph.)

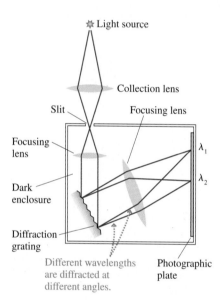

FIGURE 24.1 A diffraction spectrometer for the accurate measurement of wavelengths.

in Part VII. Our goal in this chapter, as we conclude our study of waves, is to use our knowledge of particles and waves to examine some of the experimental evidence that led to quantum physics. By doing so, we will discover the limits of the wave and particle models that we've developed.

24.1 Spectroscopy: Unlocking the Structure of Atoms

The basic discoveries of the interference and diffraction of light were made early in the 19th century. These phenomena were well understood by the end of the century, and the knowledge was used to design practical tools for measuring wavelengths with great accuracy. The primary instrument for measuring the wavelengths of light is a **spectrometer,** such as the one shown in Figure 24.1. The heart of a spectrometer is a diffraction grating that diffracts different wavelengths of light at different angles. A lens then focuses the interference fringes onto a *photographic plate* or (more likely today) an electronic array detector.

Each wavelength in the light is focused to a different position on the plate. The distinctive pattern of wavelengths emitted by a source of light and recorded on the detector is called the **spectrum** of the light. Spectroscopists discovered very early that there are two types of spectra, continuous spectra and discrete spectra:

■ Hot, self-luminous objects, such as the sun or an incandescent light bulb, emit a *continuous spectrum* in which a rainbow is formed by light being emitted at every possible wavelength.

■ In contrast, the light emitted by a gas discharge tube (such as those used to make neon signs) contains only certain discrete, individual wavelengths. Such a spectrum is called a **discrete spectrum.**

Figure 24.2 shows examples of spectra as they would appear on the photographic plate of a spectrometer. Each bright line, called a **spectral line,** represents *one* specific wavelength present in the light emitted by the source. A discrete spectrum is sometimes called a **line spectrum** because of its appearance on the plate. You can see that a neon light has its familiar reddish-orange color because nearly all of the wavelengths emitted by neon atoms fall within the wavelength range 600–700 nm that we perceive as orange and red.

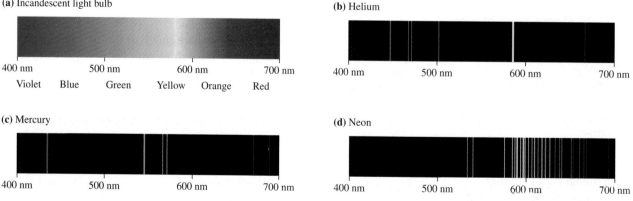

FIGURE 24.2 Examples of spectra in the visible wavelength range 400–700 nm.

Two important conclusions had been established by the end of the 19th century:

1. The light emitted by atoms in a gas discharge tube has a discrete spectrum.
2. Every element in the periodic table has its own unique spectrum.

The fact that each element emits a unique spectrum means that atomic spectra can be used as "fingerprints" to identify elements. Consequently, atomic spectroscopy is the basis of many contemporary technologies for analyzing the composition of unknown materials, monitoring air pollutants, and studying the atmospheres of the earth and other planets.

You know from chemistry that an element's *atomic number* specifies the number of protons and electrons within an atom. Hydrogen, with atomic number 1, has one electron and one proton while neon, at atomic number 10, has 10 electrons and 10 protons. It was soon recognized that an atom's internal structure determines the wavelengths the atom emits. If we only knew how to "decode" an element's spectrum, we would be able to determine the trajectories of the electrons within the atom.

Despite heroic attempts by some of the best scientists of the late 19th century, Newtonian mechanics and the (then) new theory of electromagnetism were completely unable to provide an explanation of atomic spectra or atomic structure. Not only did they fail to predict why one element's spectrum should differ from another, these classical theories predicted that atomic electrons should spiral into the nucleus, destroying the atoms and the universe in a small fraction of a second! This prediction is obviously incorrect.

Physics from the time of Newton through the mid-19th century had been spectacularly successful. But the physics of particles and waves was completely unable to explain the puzzle of discrete spectra. The first hint of a new direction in which to turn was made in 1885 by a Swiss school teacher named Johann Balmer.

Balmer and the Hydrogen Atom

Balmer was intrigued by the spectrum of hydrogen. Hydrogen is the simplest atom, with a single electron orbiting a proton, and it also has the simplest atomic spectrum. The *visible spectrum* of hydrogen, between 400 nm and 700 nm, consists of a mere four spectral lines. The wavelengths are given in Table 24.1. Physicists felt certain that such a simple spectrum must have a simple and straightforward explanation.

In 1885, Johann Balmer found by trial and error that the four wavelengths in the visible spectrum of hydrogen could be represented by the simple formula

$$\lambda = \frac{91.18 \text{ nm}}{\left(\frac{1}{2^2} - \frac{1}{n^2}\right)} \qquad n = 3, 4, 5, 6 \qquad (24.1)$$

Balmer's formula was able to reproduce the measured wavelengths with much better than 0.1% accuracy. Not only was his formula accurate, it was *simple,* in keeping with expectations that the hydrogen spectrum should have a simple explanation.

Balmer knew only the four *visible* wavelengths shown in Table 24.1, but an obvious question to ask was whether Equation 24.1 also predicts wavelengths for $n = 7$, 8, 9, and so on. The prediction for $n = 7$ is $\lambda = 397.1$ nm, an ultraviolet wavelength. Spectroscopists were just beginning to extend their craft into the ultraviolet and infrared regions of the spectrum, and it was soon confirmed that Balmer's formula does, indeed, work for *all* values of n.

Balmer's formula predicts a *series* of spectral lines of gradually decreasing wavelength, converging to the *series limit* wavelength of 364.7 nm as $n \to \infty$. Although there are an infinite number of spectral lines in this series, their intensities rapidly get weaker as n increases until, for large values of n, they blur together and cannot be resolved. This series of spectral lines is now called the **Balmer series.** Figure 24.3 shows a photograph of the Balmer series of hydrogen in which the series limit is quite obvious.

Activ
Physics ONLINE   18.2

TABLE 24.1 Wavelengths of visible lines in the hydrogen spectrum*

656.46 nm

486.27 nm

434.17 nm

410.29 nm

*Wavelengths in vacuum.

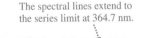
The spectral lines extend to the series limit at 364.7 nm.

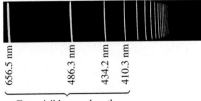

Four visible wavelengths known to Balmer

FIGURE 24.3 The Balmer series of hydrogen as seen on the photographic plate of a spectrometer.

With the success of Balmer's formula, it was natural to ask what happens if the 2^2 in Equation 24.1 is changed to 1^2 or 3^2 or m^2. It was easy to calculate that all spectral lines in the series with 1^2, if they existed, would have fairly extreme ultraviolet wavelengths while all those in the series with 3^2 would be in the infrared. Spectroscopists accepted the challenge and went to work developing the techniques for infrared and ultraviolet spectroscopy.

The $m = 1$ series was discovered by Theodore Lyman and is called the Lyman series, and the $m = 3$ series, found by Louis Paschen, is called the Paschen series. They provided confirmation, beyond any doubt, that Balmer's formula could be generalized to

$$\lambda = \frac{91.18 \text{ nm}}{\left(\dfrac{1}{m^2} - \dfrac{1}{n^2}\right)} \quad \begin{cases} m = 1 & \text{Lyman series} \\ m = 2 & \text{Balmer series} \\ m = 3 & \text{Paschen series} \\ \vdots \end{cases} \tag{24.2}$$

$$n = m + 1, m + 2, \ldots$$

As spectroscopists acquired ever more data, it became increasingly clear that Equation 24.2 could predict *every* line in the hydrogen spectrum, from the extreme ultraviolet to the far infrared.

Surely Balmer's success was not a mere coincidence. There must be some underlying meaning to his formula. But what? Balmer did not present a *theory*. He simply said, "Here's a formula that accurately calculates the wavelengths in the hydrogen spectrum." In effect, Balmer's formula was a challenge. Any successful theory of atoms must be able to *derive* Equation 24.2 from the basic laws and principles of the theory. It was 30 years before a theory was proposed that could meet this challenge.

It is particularly striking that Equation 24.2 depends on two *integers*. Hydrogen atoms simply do not emit wavelengths for $m = 1.6$ or for $n = 3.4$. This must tell us *something* important about the structure of the hydrogen atom. Newtonian mechanics does not deal in such "discrete" quantities. Masses, forces, velocities, and energies can take on any value at all; they are not restricted to having only some values but not others.

However, we have seen one exception to this: standing waves. Standing waves exist only for certain frequencies and wavelengths that are described by an *integer* called the mode number. Could there, somehow, be a connection between standing waves and the structure of atoms? To answer this question, we must probe yet deeper into the nature of light and matter.

24.2 X-Ray Diffraction

In 1895, the German physicist Wilhelm Röntgen made a remarkable discovery. The late 19th century was a period in which the technology of vacuum tubes was being perfected, and Röntgen was studying how electrons (called *cathode rays* at the time) traveled through a vacuum. He sealed an electron-producing cathode and a metal target electrode into a vacuum tube, such as shown in Figure 24.4. A high voltage pulled electrons from the cathode and accelerated them to very high speed before they struck the target. Röntgen and others had done similar experiments previously, but one day he happened by chance to have left a sealed envelope containing film near the vacuum tube. He was later surprised to discover that the film had been exposed even though it had never been removed from the envelope. This serendipitous discovery was the beginning of the study of x rays.

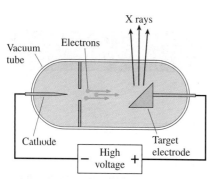

FIGURE 24.4 Röntgen's x-ray tube.

Röntgen quickly found that the vacuum tube was the source of whatever was exposing the film. But he had no idea what was coming from the tube, so he called them **x rays,** using the algebraic symbol *x* as meaning "unknown." X rays were unlike anything, particle or wave, ever discovered. Röntgen was not successful at reflecting the rays or at focusing them with a lens. He showed that they travel in straight lines, like particles, but they also pass right through most solid materials, something no known particle could do.

By the early 1900s, scientists suspected that x rays were an electromagnetic wave with a wavelength much less than that of visible light. At about the same time, scientists were first discovering that the size of an atom is ≈0.1 nm, and it was suggested that solids might consist of atoms arranged in crystalline lattices. In 1912, the German scientist Max von Laue noted that *if* x rays are waves with very short wavelengths, and *if* solids are atomic crystals with the atoms spaced about 0.1 nm apart, then x rays passing through a crystal ought to undergo diffraction from the "three-dimensional grating" of the crystal.

X-ray diffraction by crystals was soon confirmed experimentally. Measurements showed that x rays are indeed electromagnetic waves, not fundamentally different from visible light, with wavelengths in the range 0.01 nm to 10 nm.

To understand x-ray diffraction, we need to begin by looking at a crystal lattice. Figure 24.5a shows a simple cubic lattice of atoms. The crystal structure of most materials is more complex than this, but a cubic lattice will help you understand the ideas. We'll often draw just one *plane* of atoms, as in Figure 24.5b, so you'll have to visualize the three-dimensional structure of the crystal.

Suppose that a beam of x rays is incident at angle θ on the plane of atoms shown in Figure 24.6a. (Imagine the plane extending out of the page.) Most of the x rays are transmitted through the plane, because we know that x rays penetrate solids, but a small fraction of the wave may be reflected. The reflected wave obeys the law of reflection—the angle of reflection equals the angle of incidence—and the figure has been drawn accordingly.

A solid is not one plane of atoms but many parallel planes. As x rays pass through a solid, a small fraction of the wave reflects from each of the parallel planes of atoms shown in Figure 24.5b. The *net* reflection from the solid is the *superposition* of the waves reflected by each atomic plane. For most angles of incidence, the phases of the reflected waves are all different and their superposition is very near zero. In other words, as Röntgen observed, solids don't reflect x rays. However, there are a few specific angles of incidence for which the reflected waves all happened to be in phase. For these angles of incidence, the reflected waves interfere constructively to produce a strong reflection. This strong x-ray reflection at a few specific angles of incidence is called **x-ray diffraction.**

You can see from Figure 24.6b that the wave reflecting from any particular plane travels an extra distance $\Delta r = 2d\cos\theta$ before combining with the reflection from the plane immediately above it, where *d* is the spacing between the atomic planes. If $\Delta r = m\lambda$, these two waves will be in phase when they recombine. But the same geometry applies to all the planes of atoms. If the reflections from two neighboring planes are in phase, then *all* the reflections from *all* the planes are in phase and will interfere constructively to produce a strong reflection.

Consequently, x rays will reflect from the crystal when the angle of incidence θ_m satisfies

$$\Delta r = 2d\cos\theta_m = m\lambda \qquad m = 1, 2, 3, \ldots \qquad (24.3)$$

Equation 24.3 is called the **Bragg condition,** named for physicist W. L. Bragg who developed this technique for producing x-ray diffraction.

NOTE ▶ Our reasoning is very similar to the reasoning we used in Chapter 21 to understand constructive and destructive interference in thin films. ◀

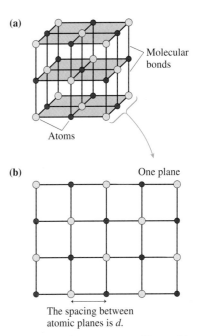

(a)

Molecular bonds

Atoms

(b)

One plane

The spacing between atomic planes is *d*.

FIGURE 24.5 Atoms arranged in a cubic lattice.

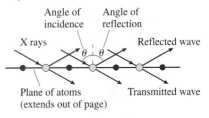

(a) X rays are transmitted and reflected at one plane of atoms.

Angle of incidence   Angle of reflection

X rays θ θ Reflected wave

Plane of atoms (extends out of page)   Transmitted wave

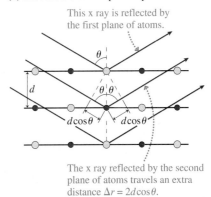

(b) The reflections from parallel planes interfere.

This x ray is reflected by the first plane of atoms.

θ

d

θ θ

$d\cos\theta$ $d\cos\theta$

The x ray reflected by the second plane of atoms travels an extra distance $\Delta r = 2d\cos\theta$.

FIGURE 24.6 The x-ray reflections from parallel atomic planes interfere constructively to cause strong reflections for certain angles of incidence.

X-ray diffraction is measured by mounting a crystal so that it can be rotated through a range of angles, as shown in Figure 24.7a. A graph of the reflected x-ray intensity versus angle θ is called the *x-ray diffraction spectrum,* and it contains valuable information about the structure of the crystal.

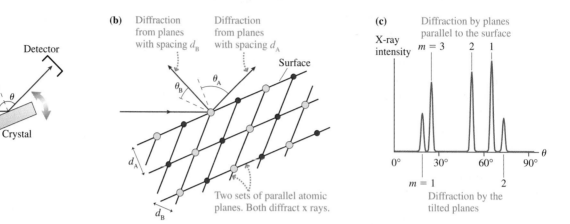

(a)

(b)

(c)

FIGURE 24.7 Producing and measuring an x-ray diffraction spectrum.

One complicating factor is that a crystal can be "sliced" into more than one set of parallel planes of atoms. Figure 24.7b shows a set of atomic planes with spacing d_A and another set of planes with spacing $d_B = d_A/\sqrt{2}$. The planes parallel to the surface cause diffraction if θ_A satisfies the Bragg condition for spacing d_A. Independently, the planes tilted at 45° cause diffraction if θ_B satisfies the Bragg condition for spacing d_B.

Figure 24.7c shows a simulated x-ray diffraction spectrum for a cubic lattice with atomic spacing $d_1 = 0.20$ nm and an x-ray wavelength $\lambda = 0.12$ nm. These are typical values. Real x-ray diffraction spectra are usually more complicated than this spectrum, but such spectra contain information with which scientists can deduce the crystalline structure of the solid.

Notice that the experimentally measured angle θ, which is measured from the surface of the crystal, is angle θ_A for the planes parallel to the surface. The experimental angle is *not* the same as angle θ_B, so it takes a little geometry to match the measured angles to the angles at which the tilted planes cause diffraction. The details will be left for a homework problem.

A modern x-ray tube that might be used for medical or dental x rays.

EXAMPLE 24.1 Analyzing x-ray diffraction

X rays with a wavelength of 0.105 nm are diffracted by a crystal. Diffraction maxima are observed at angles 31.6° and 55.4° and at no angles between these two. What is the spacing between the atomic planes causing this diffraction?

MODEL The angles must satisfy the Bragg condition. We don't know the values of m, but they are two consecutive values. Notice that θ_m *decreases* as m increases, so 31.6° corresponds to the larger value of m.

SOLVE d and λ are the same for both diffractions, so we can use the Bragg condition to find

$$\frac{m+1}{m} = \frac{\cos 31.6°}{\cos 55.4°} = 1.50 = \frac{3}{2}$$

Thus 55.4° is the second-order diffraction and 31.6° is the third-order diffraction. With this information we can use the Bragg condition again to find

$$d = \frac{2\lambda}{2\cos\theta_2} = \frac{0.105 \text{ nm}}{\cos 55.4°} = 0.185 \text{ nm}$$

ASSESS This is a reasonable value for the atomic spacing in a crystal.

Although the Bragg procedure is straightforward, most practical x-ray diffraction studies look at the diffraction of x rays that are *transmitted* through a crystal. Figure 24.8a shows a typical experimental arrangement. An x-ray tube generates several x-ray wavelengths, so Bragg diffraction is first used to select just one of these wavelengths by rotating a crystal to an angle meeting the Bragg condition. This part of the apparatus is called an *x-ray monochromator*, a device that selects one (mono) wavelength.

The known wavelength then passes through the sample and is diffracted by the three-dimensional grating of the crystal lattice. An x-ray film behind the sample records the locations of constructive interference. Because the grating is three-dimensional, the diffraction pattern consists of bright points rather than lines or fringes. Figure 24.8b shows a typical diffraction pattern. You can see that it is quite complicated. Nonetheless, crystallographers have developed many powerful analysis tools for deciphering such patterns. These techniques are computationally very intense, but modern supercomputers have made such analyses routine.

Today, x-ray diffraction is an essential tool for studying the atomic and molecular structure of solids. The most important properties of solids—their strength, chemical properties, ability to be cut or welded, optical properties, and so on—are consequences of their crystal structure. Modern engineering could not exist without the knowledge of materials gained through x-ray diffraction. Similarly, x-ray diffraction was used to deduce the double-helix structure of DNA, and it continues to elucidate the structure of biological molecules such as proteins. The techniques of x-ray diffraction are likely to become even more important in the future as physicists develop new superconducting materials, molecular biologists produce "designer drugs," and engineers design atomic-size nanostructures.

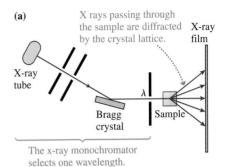

(a)

X rays passing through the sample are diffracted by the crystal lattice.

X-ray tube

X-ray film

Bragg crystal

λ

Sample

The x-ray monochromator selects one wavelength.

(b) X-ray diffraction pattern for niobium diboride

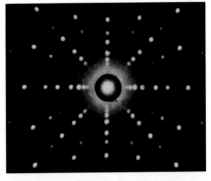

FIGURE 24.8 Using x-ray diffraction to study the atomic structure of a sample.

STOP TO THINK 24.1 The first-order diffraction of monochromatic x rays from crystal A occurs at an angle of 20°. The first-order diffraction of the same x rays from crystal B occurs at 30°. Which crystal has the larger atomic spacing?

24.3 Photons

Figure 24.9 shows three photographs made with a camera in which the film has been replaced by a special high-sensitivity detector. A correct exposure, at the right, shows a perfectly normal photograph of a woman. But with very faint illumination (left), the picture is *not* just a dim version of the properly exposed photo. Instead, it is a collection of dots. A few points on the detector have registered the presence of light, but most have not. As the illumination increases, the density of these dots increases until the dots form a full picture.

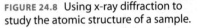

The photo at very low light levels shows individual points, as if particles are arriving at the detector.

The particle-like behavior is not noticeable at higher light levels.

Increasing light intensity

FIGURE 24.9 Photographs made with an increasing level of light intensity.

This is not what we expected. If light is a wave, reducing its intensity should cause the picture to grow dimmer and dimmer until disappearing, but the entire picture would remain present. It should be like turning down the volume on your stereo until you can no longer hear the sound. Instead, the left photograph in Figure 24.9 looks as if someone randomly threw "pieces" of light at the detector, causing full exposure at a few *discrete* points but no exposure at others.

If we did not know that light is a wave, we would interpret the results of this experiment as evidence that light is a stream of some type of particle-like object. If these particles arrive frequently enough, they overwhelm the detector and it senses a steady "river" instead of the individual particles in the stream. Only at very low intensities do we become aware of the individual particles.

Double-Slit Interference Revisited

The particle-like behavior of light seen in Figure 24.9 was apparent only for very low-intensity light. Let's return to the experiment that showed most dramatically the wave nature of light—Young's double-slit interference experiment—and lower the light intensity by inserting filters between the light source and the slits. We cannot expect to see the interference fringes by eye for such a low intensity, so we will replace the viewing screen with the same detector used to make the photographs of Figure 24.9.

What might we predict for the outcome of this experiment? If light is a wave, there is no reason to think that the nature of the interference fringes will change. The detector should continue to show alternating light and dark bands that become dimmer and dimmer until they vanish.

Figure 24.10 shows the outcome of such an experiment at three low but increasing light levels. Contrary to our prediction, the detector does not show bands at all. Instead, it shows dots like those seen in Figure 24.9. The detector is registering particle-like objects. They arrive one-by-one, and they are localized at a specific point on the detector. This is particle-like behavior, not wave-like behavior. (Waves, you will recall, are not localized at a specific point in space.) But these dots of light are not entirely random. They are grouped into bands at *exactly* the positions where we expected to see bright constructive interference fringes.

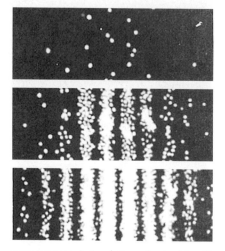

FIGURE 24.10 A simulation of a double-slit interference experiment with very low but increasing levels of light.

The Photon Model of Light

Figures 24.9 and 24.10 are our first evidence of the particle-like nature of light. These particle-like components of light are called **photons.** The concept of the photon was introduced by Albert Einstein to explain an experiment called the photoelectric effect, an experiment we will investigate in detail in Part VII.

The **photon model** of light consists of three basic postulates:

1. Light consists of discrete, massless units called photons. A photon travels in vacuum at the speed of light, 3.00×10^8 m/s.
2. Each photon has energy

$$E_{\text{photon}} = hf \tag{24.4}$$

where f is the frequency of the light and h is a *universal constant* called **Planck's constant.** The value of Planck's constant is

$$h = 6.63 \times 10^{-34} \text{ J s}$$

In other words, the light comes in discrete "chunks" of energy hf.

3. The superposition of a sufficiently large number of photons has the characteristics of a classical light wave.

EXAMPLE 24.2 The energy of a photon

550 nm is the average wavelength of visible light.

a. What is the energy of a photon with a wavelength of 550 nm?
b. A typical incandescent light bulb emits about 1 J of visible light energy every second. Estimate the number of photons emitted per second.

SOLVE

a. The frequency of the photon is

$$f = \frac{c}{\lambda} = \frac{3.00 \times 10^8 \text{ m/s}}{550 \times 10^{-9} \text{ m}} = 5.45 \times 10^{14} \text{ Hz}$$

Equation 24.4 gives us the energy of this photon:

$$E_{photon} = hf = (6.63 \times 10^{-34} \text{ Js})(5.45 \times 10^{14} \text{ Hz})$$

$$= 3.61 \times 10^{-19} \text{ J}$$

This is an extremely small energy!

b. The photons emitted by a light bulb will span a range of energies, because the light spans a range of wavelengths, but the *average* photon energy will correspond to a wavelength near 550 nm. Thus we can estimate the number of photons in 1 J of light as

$$N \approx \frac{1 \text{ J}}{3.61 \times 10^{-19} \text{ J/photon}} \approx 3 \times 10^{18} \text{ photons}$$

A typical light bulb emits about 3×10^{18} photons every second.

ASSESS This is a staggeringly large number. It's not surprising that in our everyday life we would sense only the river and not the individual particles within the flow.

Most light sources with which you are familiar emit such vast numbers of photons that you are only aware of their wave-like superposition, just as you only notice the roar of a heavy rain on your roof and not the individual raindrops. But at extremely low intensities the light begins to appear as a stream of individual photons, like the random patter of raindrops when it is barely sprinkling. Each dot on the detector in Figures 24.9 and 24.10 signifies a point where one individual photon delivered its energy and caused a measurable signal.

Although photons are particle like, they are certainly not classical particles. Classical particles, such as Newton's corpuscles of light, would travel in straight lines through the two slits of a double-slit experiment and make just two bright areas on the detector. Instead, as Figure 24.10 shows, the *particle*-like photons seem to be landing at places where a *wave* undergoes constructive interference, thus forming the bands of dots.

Suppose that the detector in the double-slit interference experiment is 30 cm behind the slits and that the light intensity so low that only 10^6 photons arrive per second. This is experimentally quite feasible. On average, a new photon passes through the slits every 10^{-6} s. A photon moving at the speed of light travels distance $d = c\Delta t = 300$ m during 10^{-6} s. While one photon is traveling the 30 cm between the slits and the detector, the next photon is 300 m away. Or in the likely case that the light source is closer to the slits than 300 m, the next photon has not yet even been emitted by the light source! Under these conditions, only one photon at a time is passing through the double-slit apparatus.

If particle-like photons arrive at the detector in a banded pattern as a consequence of wave-like interference, as Figure 24.10 shows, but if only one photon at a time is passing through the experiment, what is it interfering with? The only possible answer is that the photon is somehow interfering *with itself*. Nothing else is present. But if each photon interferes with itself, rather than with other photons, then each photon, despite the fact that it is a particle-like object, must somehow go through *both* slits!

This all seems pretty crazy. But crazy or not, this is the way light behaves. Sometimes it exhibits particle-like behavior and sometimes it exhibits wave-like behavior. You may be expecting that we will now bring forth an "explanation" so that these observations will all "make sense." Sorry. This is simply how light really and truly behaves. The thing we call *light* is stranger and more complex than it first appeared, and there just is no way for these seemingly contradictory behaviors to make sense. We have to accept nature as it is rather than hoping that nature will conform to our expectations. Furthermore, this half-wave/half-particle behavior is not restricted to light.

STOP TO THINK 24.2 Does a photon of red light have more or less energy than a photon of blue light?

24.4 Matter Waves

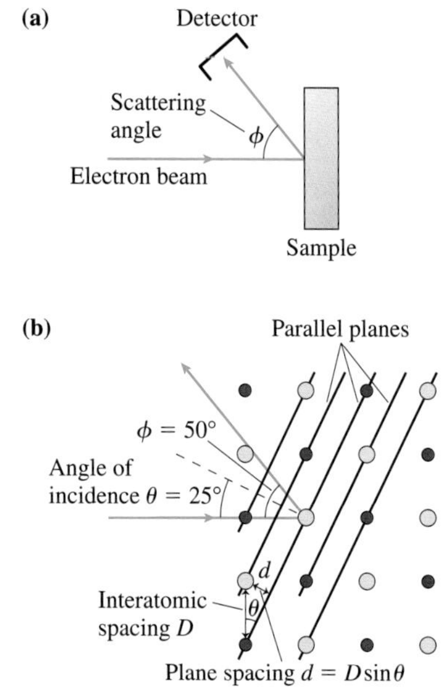

An important experiment took place in 1927 at the Bell Telephone Laboratories in New York. Two physicists, Clinton Davisson and Lester Germer, were studying how electrons scatter from the surface of metals. They had been doing similar experiments for several years, but this time they happened to use a well-crystallized piece of nickel as their target. As they rotated the electron detector around the sample, as shown in Figure 24.11a, they discovered that the intensity of the scattered electron beam exhibited clear minima and maxima.

Notice that Davisson and Germer's experiment was very similar to the Bragg x-ray diffraction experiment that was shown in Figure 24.7a. And the scattered-electron intensity they observed was not unlike the x-ray intensity pattern shown in Figure 24.7c. Although we "know" that electrons are material particles, completely unlike light waves, suppose we were to analyze the Davisson-Germer experiment *as if* electrons were waves undergoing Bragg diffraction.

Davisson and Germer found that electrons incident normal to the crystal face at a speed of 4.35×10^6 m/s scattered at $\phi = 50°$. You can see in Figure 24.11b that this scattering can be interpreted as a mirror-like reflection from the atomic planes that slice diagonally through the crystal. The angle of incidence on this set of planes is $\theta = \phi/2 = 25°$. This is the angle in Equation 24.3, $2d\cos\theta_m = m\lambda$, the Bragg condition for diffraction.

You can also see that the spacing d between the atomic planes is related to the atomic spacing D by

$$d = D\sin\theta \tag{24.5}$$

Equation 24.5 allows us to write the Bragg condition in terms of the atomic spacing D, rather then the plane spacing d, as

$$2(D\sin\theta_m)\cos\theta_m = D(2\sin\theta_m\cos\theta_m) = D\sin(2\theta_m) = m\lambda \tag{24.6}$$

From x-ray diffraction, the atomic spacing of nickel was already known to be $D = 0.215$ nm. If we combine this value of D with the measured angle $\theta = 25°$, and if we assume $m = 1$, then we find that the "electron wavelength" is

$$\lambda = D\sin(2\theta) = 0.165 \text{ nm} \tag{24.7}$$

This seems like a pointless exercise. Yes, electrons reflect from a nickel surface with a scattering angle of 50°. But electrons are particles of matter, so there must be some explanation in terms of the collision of particles with the atoms at the surface of the crystal. Right? Nonetheless, Davisson and Germer searched for, and quickly found, 20 other reflections obeying the Bragg condition for *exactly* the same "wavelength" of 0.165 nm.

These results could not be a coincidence. Electrons, particles of matter with mass, were somehow, in some way, being *diffracted* by the grating of a crystal. Particles of matter, for the first time ever, were being observed to have wave-like properties!

(a)

Detector

Scattering angle ϕ

Electron beam

Sample

(b)

Parallel planes

$\phi = 50°$

Angle of incidence $\theta = 25°$

Interatomic spacing D

Plane spacing $d = D\sin\theta$

FIGURE 24.11 The Davisson-Germer experiment to study electrons scattered from metal surfaces.

The de Broglie Wavelength

Three years earlier, in 1924, a French graduate student named Louis-Victor de Broglie (Figure 24.12) was puzzling over the growing evidence that light seemed to have both wave-like and particle-like properties. Sometimes light acted like a classical wave, exhibiting interference and diffraction. Yet at other times, light seemed to come in small, localized pieces like a particle. Einstein had just won the Nobel prize in 1921 for his explanation of the photoelectric effect in terms of particle-like photons of light.

If light, something that we generally think of as a wave, can act like a particle, then it occurred to de Broglie that perhaps some object we generally think of as a particle would, in the right conditions, act like a wave. What are the most "particle-like" entities we can think of? Very likely electrons and protons, the basic building blocks of matter. Can an electron or a proton act like a wave? If they did so, how would we know? What behavior would they exhibit that is wave-like? And what is the "wavelength" of an electron—if it has one?

De Broglie postulated that a particle of mass m and momentum $p = mv$ has a wavelength

$$\lambda = \frac{h}{p} \tag{24.8}$$

FIGURE 24.12 Louis-Victor de Broglie.

where h is Planck's constant. This wavelength for material particles is now called the **de Broglie wavelength.** It depends *inversely* on the particle's momentum, so the largest wave effects will occur for particles having the smallest momentum.

What led de Broglie to this postulate? The constant that we now symbolize as h was first introduced into physics in the year 1900 by the German physicist Max Planck as part of an explanation of how incandescent objects emit continuous spectra. Planck's explanation was not widely accepted until 1905, when Einstein showed that the photoelectric effect could be understood if the energy E of a photon of light is related to its frequency f by $E_{\text{photon}} = hf$.

It was this relationship of energy to frequency that intrigued de Broglie. He reasoned that if matter has wave-like properties, it should also obey Einstein's $E = hf$. But he also knew that the kinetic energy of a particle of mass m is related to its momentum by

$$E = \frac{1}{2}mv^2 = \frac{1}{2}m\left(\frac{p}{m}\right)^2 = \frac{p^2}{2m} \tag{24.9}$$

What relationship between momentum and wavelength would allow these two statements about the particle's energy to be consistent with each other? The only possibility de Broglie could find was $\lambda = h/p$. The details of his reasoning, although not difficult, are not important to us. Our goal, instead, is to understand the experimental evidence for, and some of the implications of, de Broglie's bold and imaginative suggestion.

It is worth noting that there was absolutely *no* evidence for matter waves in 1924. Even so, de Broglie must have reasoned, perhaps the evidence was lacking only because no one had looked in the right places or used the right equipment and techniques. If Equation 24.8 is correct, what evidence would you expect to see? You know that the most obvious characteristic of waves is their ability to exhibit interference and diffraction, but you have also learned that diffraction effects are not easily observable unless the opening through which a wave passes is comparable in size to the wavelength. There is no obvious spreading when a wave passes through an opening of size $a \gg \lambda$. What wavelengths do material particles have, and is it likely that anyone would have seen their diffraction before 1924?

EXAMPLE 24.3 The de Broglie wavelength of a smoke particle

One of the smallest macroscopic particles we could imagine using for an experiment would be a very small smoke or soot particle. These are $\approx 1 \ \mu m$ in diameter, too small to see with the naked eye and just barely at the limits of resolution of a microscope. A particle this size has mass $m \approx 10^{-18}$ kg. Estimate the de Broglie wavelength for a 1 μm diameter particle moving at the very slow speed of 1 mm/s.

SOLVE The particle's momentum is $p = mv \approx 10^{-21}$ kg m/s. The de Broglie wavelength of a particle with this momentum is

$$\lambda = \frac{h}{p} \approx 7 \times 10^{-13} \ m$$

ASSESS This wavelength is $\approx 1\%$ the size of an atom. We can't shoot a 1-μm-diameter particle though an atom-size hole, so we certainly don't expect to see any wave-like behavior. And if a 1 μm particle has a wavelength this small, the wavelength of a baseball must be smaller by many factors of 10. It is thus little wonder, if de Broglie's suggestion is correct, that we do not see macroscopic objects exhibiting wave-like behavior.

EXAMPLE 24.4 The de Broglie wavelength of an electron

Find the de Broglie wavelength of an electron with a speed of 4.35×10^6 m/s, the speed used in the Davisson-Germer experiment.

SOLVE The mass of an electron is 9.11×10^{-31} kg. Its de Broglie wavelength at this speed is

$$\lambda = \frac{h}{p} = \frac{h}{mv} = 0.167 \ nm$$

ASSESS This result is in near-perfect agreement with Davisson and Germer's experimentally determined wavelength of 0.165 nm! Electrons moving with speeds in this range have de Broglie wavelengths very similar to those of x rays. These wavelengths are exactly the right size to be diffracted by atomic crystals.

Davisson and Germer, who won the Nobel prize for their demonstration of the wave nature of electrons, had not set out to perform a breakthrough experiment. They were simply continuing research that had started years earlier, and they had never heard of de Broglie at the time they found unexpected and unexplainable results. However, being open-minded enough to seek out the advice and opinion of others, they learned that they might be able to demonstrate electron diffraction. A large element of chance and luck was involved; they just happened to be doing the right experiments at the right time. But their careful thought and study had also prepared them to recognize a unique opportunity when it came along. It was their willingness to give a fair test to a really crazy idea—that electrons might be waves!—that earned them a place in science history.

The Interference and Diffraction of Matter

17.5 Activ Physics

Further evidence in support of de Broglie's hypothesis was soon forthcoming. The English physicist G. P. Thompson performed a diffraction experiment with an electron beam transmitted *through* a crystal, an experiment exactly equivalent to Figure 24.8 for x-ray diffraction. Figures 24.13a and 24.13b show the diffraction patterns produced by x rays and electrons passing through an aluminum-foil target. (The foil is not a single crystal but, instead, thousands of tiny crystal grains at random orientations. As a consequence, the single-crystal diffraction spots of Figure 24.8b get rotated about the axis and form concentric diffraction circles.) The primary observation to make from Figure 24.13 is that **electrons diffract exactly like x rays.**

(a) X-ray diffraction pattern

(b) Electron diffraction pattern

(c) Neutron diffraction pattern

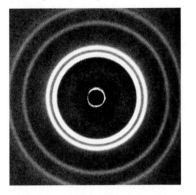

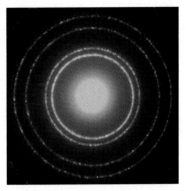

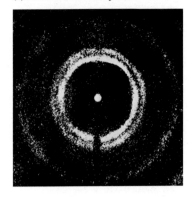

FIGURE 24.13 The diffraction patterns produced by x rays, electrons, and neutrons passing through an aluminum-foil target.

Later experiments demonstrated that de Broglie's hypothesis applies to other material particles as well. Neutrons have a much larger mass than electrons, which tends to decrease their de Broglie wavelength, but it is possible to generate very slow neutrons. The much smaller speed compensates for the heavier mass, so neutron wavelengths can be comparable to electron wavelengths. Figure 24.13c shows a neutron diffraction pattern. It is similar to the x ray and electron diffraction patterns, although of lower quality because neutrons are harder to detect. A neutron, too, is a matter wave. In fact, in recent years it has become possible to observe the interference and diffraction of entire atoms!

The classic test of "waviness" is Young's double-slit interference experiment. If an electron, or other material object, has wave-like properties, it should exhibit interference when passing through two slits. Does it? This experiment is not easy to do because the spacing between the two slits has to be very tiny. The technical challenges of such an experiment could not be met until around 1960, when it became possible to produce slits in a thin foil with a spacing of $\approx 2 \ \mu\text{m}$. Even then, various technical reasons required the electrons to have much higher velocities than Davisson and Germer used, reducing their de Broglie wavelength to ≈ 0.005 nm. This rather significant discrepancy between the wavelength and the slit spacing is equivalent to an optical double-slit experiment with a slit spacing of 20 cm. Nonetheless, the experiment was performed, and Figure 24.14a shows the highly enlarged electron pattern that was detected. Amazing as it seems, electrons, one of the basic building blocks of matter, produce interference fringes after passing through a double slit.

(a) Double-slit interference of electrons

(b) Double-slit interference of neutrons

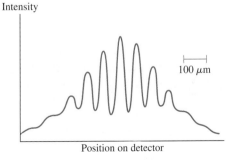

FIGURE 24.14 Double-slit interference patterns of electrons and neutrons.

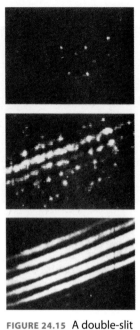

FIGURE 24.15 A double-slit interference pattern of electrons is built up electron by electron as they arrive at the detector.

Later, during the 1970s and 1980s, techniques were developed for observing the double-slit interference of neutrons. Figure 24.14b shows the pattern recorded when neutrons passed through two slits separated by 0.10 mm. The characteristic interference fringes are readily observed, despite the much larger mass of the neutron.

Figure 24.15 shows an electron double-slit experiment in which the intensity of the electron beam was reduced to only a few electrons per second. You can see that each electron is detected on the screen as a *particle*, a localized dot where the electron hits, but that the pattern of dots is the interference pattern of a *wave* with wavelength $\lambda = h/p$. Compare this picture to Figure 24.10 for photons. (Note that Figure 24.10 was a simulation, but Figure 24.15 is a photograph from a real experiment.) In both cases, electrons and photons, we see a combination of both wave-like and particle-like behaviors.

We noted earlier that each photon must in some sense interfere with itself. The same is true for electrons. If only a few electrons arrive per second, then only one electron at a time is in the region of the slits and the screen. Each electron somehow goes through both slits, has a wave-like interference with itself, but is finally detected at the screen as a particle-like dot.

NOTE ▶ We are *not* saying that photons and electrons are the same thing. We are saying that light and electrons are found to share both wave-like and particle-like properties, so under similar experimental conditions we can expect to see similar behavior. Nonetheless, electrons are matter. They are particles with mass and charge that obey $\lambda = h/p$. Photons have no mass, no charge, and obey $\lambda = c/f$. There are many situations in which the behaviors of electrons and photons are quite distinct. ◀

STOP TO THINK 24.3   A proton, an electron, and an oxygen atom each pass at the same speed through a 1-μm-wide slit. Which will produce a wider diffraction pattern on a detector behind the slit?

a. The proton.
b. The electron.
c. The oxygen atom.
d. All three will be the same.
e. None of them will produce a diffraction pattern.

24.5 Energy Is Quantized

De Broglie hypothesized that material particles have wave-like properties, and you've now seen experimental evidence that this must be true. This final section will explore one of the most important implications of the wave-like nature of matter.

You learned in Chapter 21 that waves confined between two boundaries form standing waves. Wave reflections cause the region between the two boundaries to be filled with waves traveling in both directions, and the superposition of two oppositely directed waves produces a standing wave.

To see how this applies to matter, let's consider what physicists call "a particle in a box." Figure 24.16 shows a particle of mass m confined inside a rigid box of length L. For simplicity, we'll consider only the one-dimensional motion parallel to the length of the box. Furthermore, we'll assume that collisions with the ends of the box are perfectly elastic, with no loss of kinetic energy.

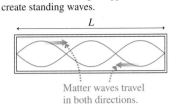

(a) A classical particle of mass m bounces back and forth between the ends.

(b) Matter waves moving in opposite directions create standing waves.

Matter waves travel in both directions.

FIGURE 24.16 A particle of mass m confined in a box of length L.

Figure 24.16a shows a classical particle, such as a ball or a dust particle, in the box. This particle will simply bounce back and forth at constant speed. But if particles have wave-like properties, perhaps we should consider a *wave* reflecting back and forth from the ends of the box. The reflections will create the standing wave shown in Figure 24.16b. This standing wave is analogous to the standing wave on a string that is tied at both ends.

In Chapter 21, we found that the wavelength of a standing wave is related to the length L of the confining region by

$$\lambda_n = \frac{2L}{n} \qquad n = 1, 2, 3, 4, \ldots \tag{24.10}$$

The particle must also satisfy the de Broglie condition $\lambda = h/p$. Equating these two expressions for the wavelength gives

$$\frac{h}{p} = \frac{2L}{n} \tag{24.11}$$

Solving Equation 24.11 for the particle's momentum p, we find

$$p_n = n\left(\frac{h}{2L}\right) \qquad n = 1, 2, 3, 4, \ldots \tag{24.12}$$

Equation 24.12 is a most surprising result. It appears that the momentum of a wave-like particle can have only those *discrete* values given by Equation 24.12. Newtonian physics places no restrictions on the value of the momentum, hence Equation 24.12 represents a clear break with Newtonian physics.

The particle's energy, which is entirely kinetic energy, is related to its momentum by

$$E = \frac{1}{2}mv^2 = \frac{p^2}{2m} \tag{24.13}$$

If we use Equation 24.12 for the momentum, we find that the particle's energy is restricted to the discrete values

$$E_n = \frac{1}{2m}\left(\frac{hn}{2L}\right)^2 = \frac{h^2}{8mL^2}n^2 \qquad n = 1, 2, 3, 4, \ldots \tag{24.14}$$

This conclusion is one of the most profound discoveries of physics. Because of the wave nature of matter, which has ample experimental confirmation, **a confined particle can have only certain energies.** It is simply not possible for the particle to exist in the box with any energy other than one of the values given by Equation 24.14.

This result, that a confined particle can have only discrete values of energy, is called the **quantization** of energy. More informally, we say that energy is *quantized.* The number n is called the **quantum number,** and each value of n characterizes one **energy level** of the particle in the box.

Not only is the energy quantized, we see from Equation 24.14 that the energy of the particle in the box cannot be reduced below

$$E_1 = \frac{h^2}{8mL^2} \tag{24.15}$$

E_1 is the *least* kinetic energy a particle can have. Because $E_1 > 0$, **the particle is *always* in motion;** it cannot be made to stay at rest! These properties of a wave-like particle in a box are in stark contrast to those of a classical Newtonian particle, for which the possible energies are continuous and the minimum kinetic energy is zero. In terms of E_1, the allowed energies are

$$E_n = n^2 E_1 \tag{24.16}$$

This result is analogous to our earlier finding that standing waves can exist for only the discrete frequencies $f_n = nf_1$.

Notice that the allowed energies are inversely proportional to both m and L^2. The quantization of energy is not apparent with macroscopic objects, or else we would have known about it long ago, so both m and L have to be exceedingly small before energy quantization has any significance. This is an important observation because any new theory about matter and energy cannot be in conflict with our observations of macroscopic objects. Newtonian physics still works for baseballs.

EXAMPLE 24.5 The minimum energy of a smoke particle

What is the first allowed energy of the very small 1-μm-diameter particle of Example 24.3 if it is confined to a very small box 10 μm in length?

SOLVE This is about as small as we can easily imagine making macroscopic particles and boxes. Example 24.3 noted that such a particle has $m \approx 10^{-18}$ kg. The first allowed energy, $n = 1$, is

$$E_1 = \frac{h^2}{8mL^2} \approx 5 \times 10^{-40} \text{ J}$$

ASSESS This is an unimaginably small amount of energy. By comparison, the kinetic energy of a 1-μm-diameter particle moving at a barely perceptible speed of 1 mm/s is $K \approx 5 \times 10^{-25}$ J, a factor of 10^{15} larger. There is no way we could ever observe or measure discrete energies this small, so it is not surprising that we are unaware of energy quantization for macroscopic objects.

EXAMPLE 24.6 The minimum energy of an electron

What are the first three allowed energies of an electron confined to a 0.10-nm-long box?

SOLVE The mass of an electron is $m = 9.11 \times 10^{-31}$ kg. Thus the first allowed energy is

$$E_1 = \frac{h^2}{8mL^2} = 6.0 \times 10^{-18} \text{ J}$$

This is the lowest allowed energy. The next two allowed energies are

$$E_2 = 2^2 E_1 = 24.0 \times 10^{-18} \text{ J}$$
$$E_3 = 3^2 E_1 = 54.0 \times 10^{-18} \text{ J}$$

ASSESS An electron with energy E_1 has speed $v = 3.6 \times 10^6$ m/s, roughly 1% of the speed of light. A 0.10-nm-long box is about the size of an atom. The very large speed of an electron with the *minimum* electron energy in an atomic-size box suggests that the wave nature of electrons *is* important for the physics of atoms.

Energy quantization is simply not an issue for the physics of macroscopic objects. Newtonian physics works fine. But at the much smaller scale of atoms, the physics is very new and different. An atom is certainly more complicated than a simple one-dimensional box, but an electron is "confined" within an atom rather as a particle is in a box. Thus the electron orbits must, in some sense, be standing waves. This idea will be the key that unlocks the mystery of discrete atomic spectra. We will study the atom more carefully in Part VII, but you can see from this introduction that the quantization of energy *is* important at the microscopic scale of the atom.

These examples raise more questions than they answer. If matter is some kind of wave, what is waving? What is the medium of a matter wave? What kind of displacement does it undergo? De Broglie's hypothesis is not a *theory,* and it provides no answers to important questions such as these.

De Broglie's suggestion came nearly 40 years after Balmer's discovery, 40 years during which the atom was being explored and the failures of classical physics were becoming ever more apparent. His suggestion was the final spark, setting off a burst of activities and new ideas that led within a year to a complete and revolutionary new theory—quantum physics. We will revisit these issues later, but for now it is important to see just how far we have been able to come with our study of waves.

STOP TO THINK 24.4   A proton, an electron, and an oxygen atom are each confined in a 1-nm-long box. Rank in order, from largest to smallest, the minimum possible energies of these particles.

SUMMARY

The goal of Chapter 24 has been to explore the limits of the wave and particle models.

GENERAL PRINCIPLES

The two basic models of classical physics

The particle model

A particle is localized at one point in space.
A particle follows a well-defined trajectory.

The wave model

A wave is spread out through space.
A wave exhibits interference and diffraction.

The breakdown of classical physics

A closer look at light and matter finds that these classical models are not sufficient. Light and matter are neither particles nor waves, but exhibit characteristics of both.

IMPORTANT CONCEPTS

Light

- Exhibits interference and diffraction
  Wave-like: $c = \lambda f$
- Detected at localized positions
  Particle-like: $E = hf$
- Particle-like "chunks" of light are called photons.

Matter

- Detected at localized positions
  Particle-like: $E = \frac{1}{2}mv^2$
- Exhibits interference and diffraction
  Wave-like: $\lambda = h/p$
- The wavelength is called the **de Broglie wavelength.**

Quantization

A "particle" confined to one-dimensional box of length L sets up a standing wave with the de Broglie wavelength. Because only certain wavelengths can oscillate, only certain discrete energies are allowed:

$$E_n = \frac{h^2}{8mL^2} n^2 \qquad n = 1, 2, 3, \ldots$$

Energy is quantized into discrete levels rather than being continuous as it is in classical physics. Quantization is not important for macroscopic objects, but energy quantization plays a very large role at the atomic level.

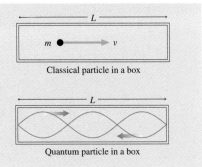

Classical particle in a box

Quantum particle in a box

APPLICATIONS

Hydrogen spectrum

The wavelengths in the spectrum of hydrogen atoms are

$$\lambda = \frac{91.18 \text{ nm}}{\left(\dfrac{1}{m^2} - \dfrac{1}{n^2}\right)} \qquad \begin{array}{l} m = 1, 2, 3, \ldots \\ n = m + 1, m + 2, \ldots \end{array}$$

The series of spectral lines with $m = 2$ is the Balmer series.

Diffraction by atomic crystals

X rays and matter particles with wavelength λ undergo strong reflections from atomic planes spaced by d when the angle of incidence satisfies the Bragg condition

$$2d\cos\theta = m\lambda \qquad m = 1, 2, 3, \ldots$$

TERMS AND NOTATION

spectrometer	line spectrum	Bragg condition	de Broglie wavelength
spectrum	Balmer series	photon	quantization
discrete spectrum	x ray	photon model	quantum number, n
spectral line	x-ray diffraction	Planck's constant, h	energy level, E_n

EXERCISES AND PROBLEMS

Data for Chapter 24: $m_{electron} = 9.11 \times 10^{-31}$ kg
$m_{proton} = m_{neutron} = 1.67 \times 10^{-27}$ kg

Exercises

Section 24.1 Spectroscopy: Unlocking the Structure of Atoms

1. What are the wavelengths of spectral lines in the Balmer series with $n = 6$, 8, and 10?
2. Show that the series limit of the Balmer series is 364.7 nm.
3. Which member of the Balmer series has wavelength 389.0 nm?

Section 24.2 X-Ray Diffraction

4. X rays with a wavelength of 0.12 nm undergo first-order diffraction from a crystal at a 68° angle of incidence. What is the angle of second-order diffraction?
5. X rays with a wavelength of 0.20 nm undergo first-order diffraction from a crystal at a 54° angle of incidence. At what angle does first-order diffraction occur for x rays with a wavelength of 0.15 nm?
6. X rays diffract from a crystal in which the spacing between atomic planes is 0.175 nm. The second-order diffraction occurs at 45.0°. What is the angle of the first-order diffraction?
7. X rays with a wavelength of 0.085 nm diffract from a crystal in which the spacing between atomic planes is 0.180 nm. How many diffraction orders are observed?

Section 24.3 Photons

8. What is the energy of a photon of visible light that has a wavelength of 500 nm?
9. What is the energy of an x-ray photon that has a wavelength of 1.0 nm?
10. What is the wavelength of a photon whose energy is twice that of a photon with a 600 nm wavelength?
11. What is the energy of 1 mol of photons that have a wavelength of 1.0 μm?

Section 24.4 Matter Waves

12. Estimate your de Broglie wavelength while walking at a speed of 1 m/s.

13. a. What is the speed of an electron with a de Broglie wavelength of 0.20 nm?
    b. What is the speed of a proton with a de Broglie wavelength of 0.20 nm?
14. What is the kinetic energy of an electron with a de Broglie wavelength of 1.0 nm?
15. a. What is the de Broglie wavelength of a 200 g baseball with a speed of 30 m/s?
    b. What is the speed of a 200 g baseball with a de Broglie wavelength of 0.20 nm?

Section 24.5 Energy Is Quantized

16. What is the smallest box in which you can confine an electron if you want to know for certain that the electron's speed is no more than 10 m/s?
17. What is the length of a box in which the minimum energy of an electron is 1.5×10^{-18} J?
18. The nucleus of an atom is 5.0 femtometers in diameter, where 1 femtometer = 1 fm = 10^{-15} m. A very simple model of the nucleus is a box in which protons are confined. Estimate the energy of a proton in the nucleus by finding the first three allowed energies of a proton in a 5.0-fm-long box.

Problems

19. a. Calculate the wavelengths of the first four members of the Lyman series in the spectrum of hydrogen.
    b. What is the series limit for the Lyman series?
    c. Light from a hydrogen discharge lamp passes through a diffraction grating and registers on a detector 1.5 m behind the grating. The $m = 1$ diffraction of the first member of the Lyman series is located 37.6 cm from the central maximum. What is the position of the second member of the Lyman series?
20. a. Calculate the wavelengths of the first four members of the Paschen series in the spectrum of hydrogen.
    b. What is the series limit for the Paschen series?
    c. Light from a hydrogen discharge passes through a diffraction grating and registers on a detector 1.5 m behind the grating. The $m = 1$ diffraction of the first member of the Paschen series is located 60.7 cm from the central maximum. What is the position of the second member of the Paschen series?

21. *Gamma rays* are photons with very high energy.
    a. What is the wavelength of a gamma-ray photon with energy 1.0×10^{-13} J?
    b. How many visible-light photons with a wavelength of 500 nm would you need to match the energy of this one gamma-ray photon?

22. A 1000 kHz AM radio station broadcasts with a power of 20 kW. How many photons does the transmitting antenna emit each second?

23. A helium-neon laser emits a light beam with a wavelength of 633 nm. The power of the laser beam is 1.0 mW.
    a. What is the energy of one photon of laser light?
    b. How many photons does the laser emit each second?

24. Example 24.2 found that a typical incandescent light bulb emits $\approx 3 \times 10^{18}$ visible-light photons per second. Your eye, when it is fully dark adapted, can barely see the light from an incandescent light bulb 10 km away. How many photons per second are incident at the image point on your retina? The diameter of a dark-adapted pupil is ≈ 6 mm.

25. X-ray photons with energies of 1.50×10^{-15} J are incident on a crystal. The spacing between the atomic planes in the crystal is 0.21 nm. At what angles of incidence will the x rays diffract from the crystal?

26. X rays with a wavelength of 0.070 nm diffract from a crystal. Two adjacent angles of x-ray diffraction are 45.6° and 21.0°. What is the distance between the atomic planes responsible for the diffraction?

27. a. Show that the Bragg condition for x-ray diffraction at normal incidence is equivalent to the condition for maximum reflectivity of a thin film.
    b. Researchers have recently learned how to fabricate thin-film coatings only a few atoms thick out of alternating layers of tungsten and boron carbide. These coatings are expected to greatly improve the x-ray telescopes used in astronomy. What are the two longest x-ray wavelengths that will reflect at normal incidence from a film with a thickness of 1.2 nm?

28. The basic idea of Bragg diffraction is not limited to x rays. One contemporary application is in optical fibers. It is sometimes useful to block one particular wavelength of light, by reflecting it, while transmitting all other wavelengths. Figure P24.28 shows that this can be done by building a short section of fiber, called a *fiber grating,* in which the index of refraction varies periodically. A small fraction of the light wave traveling through the fiber reflects from each little "bump" in the index of refraction. For most wavelengths, the reflected waves have random phases and their superposition is essentially zero. These wavelengths are transmitted through the fiber grating. If, however, the reflections are all in phase for some wavelength, that wavelength is strongly reflected and the transmitted light is strongly attenuated. Consider a fiber grating in a glass fiber ($n = 1.50$) with a spacing of 0.45 μm. What is the air wavelength of infrared light that is blocked by this fiber grating?

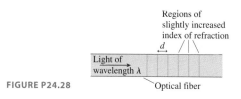

Regions of slightly increased index of refraction

Light of wavelength λ

Optical fiber

FIGURE P24.28

29. Electrons with a speed of 2.0×10^6 m/s pass through a double-slit apparatus. Interference fringes are detected with a fringe spacing of 1.5 mm.
    a. What will the fringe spacing be if the electrons are replaced by neutrons with the same speed?
    b. What speed must neutrons have to produce interference fringes with a fringe spacing of 1.5 mm?

30. Electrons pass through a 1.0-μm-wide slit with a speed of 1.5×10^6 m/s. How wide is the electron diffraction pattern on a detector 1.0 m behind the slit?

31. The double-slit neutron diffraction pattern shown in Figure 24.14b was measured 3.0 m behind two slits having a separation of 0.10 mm. From measurements that you can make *on the figure* (notice the scale on the figure), determine the speed of the neutrons.

32. In a Davisson-Germer experiment, electrons with a speed of 4.30×10^6 m/s are scattered at an angle of 60°. What is the atomic spacing D?

33. a. What are the energies of the first three energy levels of an electron confined in a one-dimensional box of length 0.70 nm?
    b. How much energy must the electron lose to move from the $n = 2$ energy level to the $n = 1$ energy level?
    c. Suppose that an electron can move from the $n = 2$ level to the $n = 1$ level by emitting a photon of light. If energy is conserved, what must the photon's wavelength be? Give your answer in nm.

34. a. What is the minimum energy of a 10 g Ping-Pong ball in a 10-cm-long box?
    b. At what speed does it have this energy?

35. What is the length of a box in which the difference between an electron's first and second allowed energies is 1.0×10^{-19} J?

36. Two adjacent allowed energies of an electron in a one-dimensional box are 3.6×10^{-19} J and 6.4×10^{-19} J. What is the length of the box?

37. a. Find an expression for the allowed velocities of a particle of mass m in a one-dimensional box of length L.
    b. What are the first three allowed velocities for an electron in a 0.20-nm-long box?

38. It can be shown that the allowed energies of a particle of mass m in a two-dimensional square box of side L are

$$E_{nm} = \frac{h^2}{8mL^2}(n^2 + m^2)$$

The energy depends on two quantum numbers, n and m, both of which must have an integer value 1, 2, 3, . . .
    a. What is the minimum energy for a particle in a two-dimensional square box of side L?
    b. What are the five lowest allowed energies? Give your values as multiples of E_{min}.

Challenge Problems

39. Let's further explore the x-ray diffraction spectrum of Figure 24.7c. This spectrum was for x rays of wavelength 0.12 nm incident on a cubic crystal with atomic spacing 0.20 nm.
    a. Calculate all the diffraction angles θ_A for diffraction from the atomic planes parallel to the surface. Do these angles agree with Figure 24.7c?

b. Calculate all the diffraction angles θ_B for diffraction from the tilted atomic planes.

c. At what measured angles θ should diffraction from the tilted atomic planes be observed? How does your answer compare to Figure 24.7c?

Hint: A second set of atomic planes is tilted in the opposite direction.

40. X rays with wavelength 0.10 nm are incident on a crystal with a hexagonal crystal structure. The x-ray diffraction spectrum is shown in Figure CP24.40. What is the atomic spacing D of this crystal?

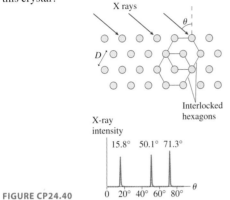

FIGURE CP24.40

41. A sound wave is easily transmitted through an open-open tube of air only if the sound's wavelength λ matches one of the possible standing waves in the tube. This is called a *standing-wave resonance*. The same idea applies to a modern semiconductor device called a *quantum-well device* in which electrons are incident on a thin layer of material that "confines" the electrons. Electrons are trapped within this layer, and it is not easy for electrons to enter or leave. However, electrons can easily flow through this layer if they are able to excite a de Broglie standing-wave resonance in the confinement layer.

a. The confinement layer in a quantum-well device is 5.0 nm thick. What are the four longest-wavelength de Broglie standing waves in this layer?

b. What are the four lowest electron speeds for which an electron current will flow through this layer?

We will study quantum well devices in more detail in Part VII. They are used to make light-emitting diodes and semiconductor lasers.

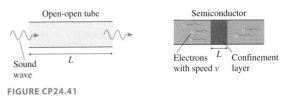

FIGURE CP24.41

42. In Chapter 22, where we studied diffraction gratings, you learned that light with wavelength λ is diffracted by a piece of matter (the grating) with a periodic structure (many slits with spacing d). Experiments done in the 1990s showed that the roles of light and matter can be reversed. That is, matter with a de Broglie wavelength λ can be diffracted by light with a periodic structure. A periodic structure of light is easily created by reflecting a laser beam back on itself to create a standing wave. The experimental challenge, which is quite difficult, is to create a "monochromatic" beam of atoms with a large de Broglie wavelength. Figure CP24.42 shows a beam of sodium atoms ($m = 3.84 \times 10^{-26}$ kg) that are all traveling with a uniform speed of 50 m/s. The atomic beam crosses a laser-beam standing wave with a wavelength of 600 nm. Assuming that the diffraction obeys the diffraction-grating equation (it does), how far will the first-order-diffracted atoms be deflected sideways on a detector 1.0 m behind the laser beam?

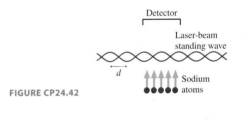

FIGURE CP24.42

STOP TO THINK ANSWERS

Stop to Think 24.1: A. The Bragg condition $2d\sin\theta_1 = \lambda$ tells us that larger values of d go with smaller values of θ_1.

Stop to Think 24.2: Less. $E = hf$, and red light, because of its longer wavelength, has a smaller frequency.

Stop to Think 24.3: b. The widest diffraction pattern occurs for the largest wavelength. The de Broglie wavelength is inversely proportional to the particle's mass.

Stop to Think 24.4: $E_{elec} > E_{proton} > E_{oxy}$. The minimum energy E_1 is inversely proportional to the particle's mass.

Waves and Optics

We end our study of waves a long distance from where we started. Who would have guessed, as we examined our first pulse on a string, that we would end up with quantum numbers? But despite the wide disparity between string waves and matter waves, a few key ideas have stayed with us throughout Part V: the principle of superposition, interference and diffraction, and standing waves. As part of your final study of waves, you should trace the influence of these ideas through the chapters of Part V.

One point we have tried to emphasize is the *unity* of wave physics. We did not need separate theories of string waves and sound waves and light waves. Instead, a few basic ideas enabled us to understand waves of all types. By focusing on similarities, we have been able to analyze sound and light as well as strings and electrons in a single part of this book.

Unfortunately, the physics of waves is not as easily summarized as the physics of particles. Newton's laws and the conservation laws are two very general sets of principles about particles, principles that allowed us to develop the powerful problem-solving strategies of Parts I and II. You probably noticed that we have not found any general problem-solving strategies for wave problems.

This is not to say that wave physics has no structure. Rather, the knowledge structure of waves and optics rests more heavily on *phenomena* than on general principles. Unlike the knowledge structure of Newtonian mechanics, which was a "pyramid of ideas," the knowledge structure of waves is a logical grouping of the major topics you studied. This is a different way of structuring knowledge, but it still provides you with a mental framework for analyzing and thinking about wave problems.

KNOWLEDGE STRUCTURE V **Waves and Optics**

ESSENTIAL CONCEPTS	Wave speed, wavelength, frequency, phase, wave front, and ray.
BASIC GOALS	What are the distinguishing features of waves?
	How does a wave travel through a medium?
	How does a medium respond to the presence of more than one wave?
	What is light and what are its properties?
GENERAL PRINCIPLES	Principle of superposition
	$v = \lambda f$ for periodic waves

Traveling Waves

- The wave speed v is a property of the medium.
- The motion of particles in the medium is distinct from the motion of the wave.
- Snapshot graphs and history graphs show the same wave from different perspectives.
- The Doppler effect of shifted frequencies is observed whenever the wave source or the detector is moving.

Standing Waves

- Standing waves are the superposition of waves moving in opposite directions.
- Nodes and antinodes are spaced by $\lambda/2$.
- Only certain discrete frequencies are allowed, depending on the boundary conditions.

Interference

- Interference is constructive if two waves are in phase: $\Delta\phi = 0, 2\pi, 4\pi, \ldots$
- Interference is destructive if two waves are out of phase: $\Delta\phi = \pi, 3\pi, 5\pi, \ldots$
- The phase difference depends on the path length difference Δr and on any phase difference of the sources.
- Beats occur when $f_1 \neq f_2$.

Light and Optics

- The wave model, used for interference and diffraction, is appropriate when apertures are comparable in size to the wavelength.
- The ray model, used for mirrors and lenses, is appropriate when apertures are much larger than the wavelength.
- Diffraction, a wave effect, limits the best possible resolution of a lens.
- The photon model (discrete chunks of energy) has both wave and particle aspects.

Matter Waves

- "Particles" are not waves, but they have wave-like aspects.
- The de Broglie wavelength is $\lambda = h/mv$.
- Standing matter waves for a confined particle lead to the quantization of energy.

Wave-Particle Duality

The various objects of classical physics are *either* particles *or* waves. There's no middle ground. Planets, projectiles, and atoms are particles or collections of particles, while sound and light are clearly waves. Particles follow trajectories given by Newton's laws; waves obey the principle of superposition, spread out, and exhibit interference. This wave-particle dichotomy seemed obvious until physicists encountered evidence that light sometimes acts like a particle and, even stranger, that matter sometimes acts like a wave. The wave-like aspects of matter continue even today to baffle physicist and nonphysicist alike, but the evidence that matter has wave-like properties is overwhelming and irrefutable.

You might at first think that light and matter are *both* a wave *and* a particle, but that idea does not work. The basic definitions of particleness and waviness are mutually exclusive. Something that is a wave—spread out, non-localized—cannot simultaneously be a discrete, localized particle. It is more profitable to conclude that light and matter are *neither* a wave *nor* a particle.

Although all the macroscopic phenomena with which we are familiar seem to be easily classified as either a wave or a particle, we have no reason to expect that phenomena outside the range of our experience have to be one or the other. And, indeed, it is at the microscopic scale of atoms and their constituents—a physical scale not directly accessible to our five senses—that the classical concepts of particles and waves turn out to be simply too limited to explain the subtleties of nature. Rather than forcing every phenomenon in nature into the mold of "wave" or "particle," we need to adopt a more open-minded approach. Let an electron be an electron and a photon a photon, and nature, rather than our assumptions, will guide us as to just what these entities are.

Although matter and light have both wave-like aspects and particle-like aspects, they show us only one face at a time. If we arrange an experiment to measure a wave-like property, such as interference, we find photons and electrons acting like waves, not particles. An experiment to look for particles will find photons and electrons acting like particles, not waves.

These two aspects of light and matter are *complementary* to each other, like a two-piece jigsaw puzzle. Neither the wave nor the particle model alone provides an adequate picture of light or matter, but taken together they provide us with a basis for understanding these elusive but most fundamental constituents of nature. This two-sided point of view is called *wave-particle duality*.

Wave-particle duality is an established and accepted principle of physics as we start the 21st century. But what does it all mean? What does wave-particle duality tell us about the nature of the universe in which we live? Scientists and nonscientists alike, for over two hundred years, felt that the clockwork universe of Newtonian physics was a fundamental description of reality. The Newtonian worldview provided a familiar and widely accepted understanding of the place and role of humans in the universe.

But wave-particle duality, along with Einstein's relativity, undermines the basic assumptions of the Newtonian worldview. The certainty and predictability of classical physics have given way to a new understanding of the universe in which chance and uncertainty play key roles. Will the 21st century see the development and acceptance of a "quantum worldview" that reflects a different understanding of nature and reality? It is interesting to speculate about what a quantum worldview would be and what implications it would have for society.

We will leave waves behind for a while as we head out in entirely new directions—but not forever. The discovery of matter waves has left us with unfinished business, and we will return in Part VII for a more in-depth study of the strange world of quantum physics. There we will find that electrons and other atomic particles are described by a completely new kind of wave called a *probability wave*.

Mathematics Review

Algebra

Using exponents:

$$a^{-x} = \frac{1}{a^x} \qquad a^x a^y = a^{(x+y)} \qquad \frac{a^x}{a^y} = a^{(x-y)} \qquad (a^x)^y = a^{xy}$$

$$a^0 = 1 \qquad a^1 = a \qquad a^{1/n} = \sqrt[n]{a}$$

Fractions:

$$\left(\frac{a}{b}\right)\left(\frac{c}{d}\right) = \frac{ac}{bd} \qquad \frac{a/b}{c/d} = \frac{ad}{bc} \qquad \frac{1}{1/a} = a$$

Logarithms:

If $a = e^x$, then $\ln(a) = x$ $\qquad \ln(e^x) = x \qquad e^{\ln(x)} = x$

$$\ln(ab) = \ln(a) + \ln(b) \qquad \ln\left(\frac{a}{b}\right) = \ln(a) - \ln(b) \qquad \ln(a^n) = n\ln(a)$$

The expression $\ln(a + b)$ cannot be simplified.

Linear equations: The graph of the equation $y = ax + b$ is a straight line. a is the slope of the graph. b is the y-intercept.

Proportionality: To say that y is proportional to x, written $y \propto x$, means that $y = ax$, where a is a constant. Proportionality is a special case of linearity. A graph of a proportional relationship is a straight line that passes through the origin. If $y \propto x$, then

$$\frac{y_1}{y_2} = \frac{x_1}{x_2}$$

Slope $a = \dfrac{\text{rise}}{\text{run}} = \dfrac{\Delta y}{\Delta x}$

y-intercept $= b$

Quadratic equation: The quadratic equation $ax^2 + bx + c = 0$ has the two solutions $x = \dfrac{-b \pm \sqrt{b^2 - 4ac}}{2a}$.

Geometry and Trigonometry

Area and volume:

Rectangle

$A = ab$

Rectangular box

$V = abc$

Triangle

$A = \frac{1}{2}ab$

Right circular cylinder

$V = \pi r^2 l$

Circle

$C = 2\pi r$

$A = \pi r^2$

Sphere

$A = 4\pi r^2$

$V = \frac{4}{3}\pi r^3$

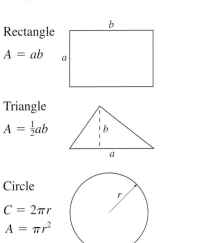
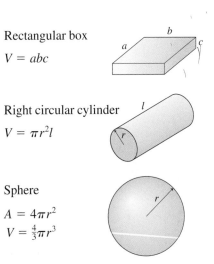

Arc length and angle: The angle θ in radians is defined as $\theta = s/r$.

The arc length that spans angle θ is $s = r\theta$.

2π rad $= 360°$

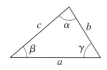

Right triangle: Pythagorean theorem $c = \sqrt{a^2 + b^2}$ or $a^2 + b^2 = c^2$

$$\sin\theta = \frac{b}{c} = \frac{\text{far side}}{\text{hypotenuse}} \qquad \theta = \sin^{-1}\left(\frac{b}{c}\right)$$

$$\cos\theta = \frac{a}{c} = \frac{\text{adjacent side}}{\text{hypotenuse}} \qquad \theta = \cos^{-1}\left(\frac{a}{c}\right)$$

$$\tan\theta = \frac{b}{a} = \frac{\text{far side}}{\text{adjacent side}} \qquad \theta = \tan^{-1}\left(\frac{b}{a}\right)$$

General triangle: $\alpha + \beta + \gamma = 180° = \pi$ rad

Law of cosines $c^2 = a^2 + b^2 - 2ab\cos\gamma$

Identities:

$$\tan\alpha = \frac{\sin\alpha}{\cos\alpha} \qquad\qquad \sin^2\alpha + \cos^2\alpha = 1$$

$$\sin(-\alpha) = -\sin\alpha \qquad\qquad \cos(-\alpha) = \cos\alpha$$

$$\sin(\alpha \pm \beta) = \sin\alpha\cos\beta \pm \cos\alpha\sin\beta \qquad \cos(\alpha \pm \beta) = \cos\alpha\cos\beta \mp \sin\alpha\sin\beta$$

$$\sin(2\alpha) = 2\sin\alpha\cos\alpha \qquad\qquad \cos(2\alpha) = \cos^2\alpha - \sin^2\alpha$$

$$\sin(\alpha \pm \pi/2) = \pm\cos\alpha \qquad\qquad \cos(\alpha \pm \pi/2) = \mp\sin\alpha$$

$$\sin(\alpha \pm \pi) = -\sin\alpha \qquad\qquad \cos(\alpha \pm \pi) = -\cos\alpha$$

Expansions and Approximations

Binomial expansion: $(1 + x)^n = 1 + nx + \dfrac{n(n-1)}{2}x^2 + \ldots$

Binomial approximation: $(1 + x)^n \approx 1 + nx$ if $x \ll 1$

Trigonometric expansions: $\sin\alpha = \alpha - \dfrac{\alpha^3}{3!} + \dfrac{\alpha^5}{5!} - \dfrac{\alpha^7}{7!} + \ldots$ for α in rad

$\cos\alpha = 1 - \dfrac{\alpha^2}{2!} + \dfrac{\alpha^4}{4!} - \dfrac{\alpha^6}{6!} + \ldots$ for α in rad

Small-angle approximation: If $\alpha \ll 1$ rad, then $\sin\alpha \approx \tan\alpha \approx \alpha$ and $\cos\alpha \approx 1$.

The small-angle approximation is excellent for $\alpha < 5°$ (≈ 0.1 rad) and generally acceptable up to $\alpha \approx 10°$.

Periodic Table of Elements

Key:
27 Co 58.9 — Atomic number / Symbol / Atomic mass

Main table

Period	Group 1	2	Transition elements →											13	14	15	16	17	18
1	1 H 1.0																		2 He 4.0
2	3 Li 6.9	4 Be 9.0												5 B 10.8	6 C 12.0	7 N 14.0	8 O 16.0	9 F 19.0	10 Ne 20.2
3	11 Na 23.0	12 Mg 24.3												13 Al 27.0	14 Si 28.1	15 P 31.0	16 S 32.1	17 Cl 35.5	18 Ar 39.9
4	19 K 39.1	20 Ca 40.1	21 Sc 45.0	22 Ti 47.9	23 V 50.9	24 Cr 52.0	25 Mn 54.9	26 Fe 55.8	27 Co 58.9	28 Ni 58.7	29 Cu 63.5	30 Zn 65.4		31 Ga 69.7	32 Ge 72.6	33 As 74.9	34 Se 79.0	35 Br 79.9	36 Kr 83.8
5	37 Rb 85.5	38 Sr 87.6	39 Y 88.9	40 Zr 91.2	41 Nb 92.9	42 Mo 95.9	43 Tc 96.9	44 Ru 101.1	45 Rh 102.9	46 Pd 106.4	47 Ag 107.9	48 Cd 112.4		49 In 114.8	50 Sn 118.7	51 Sb 121.8	52 Te 127.6	53 I 126.9	54 Xe 131.3
6	55 Cs 132.9	56 Ba 137.3	57 La 138.9	72 Hf 178.5	73 Ta 180.9	74 W 183.9	75 Re 186.2	76 Os 190.2	77 Ir 192.2	78 Pt 195.1	79 Au 197.0	80 Hg 200.6		81 Tl 204.4	82 Pb 207.2	83 Bi 209.0	84 Po 209.0	85 At 210.0	86 Rn 222.0
7	87 Fr 223.0	88 Ra 226.0	89 Ac 227.0	104 Rf 261	105 Db 262	106 Sg 263	107 Bh 264	108 Hs 269	109 Mt 268	110 Ds 271	111 272	112 285							

Inner transition elements

Lanthanides 6:

58 Ce 140.1	59 Pr 140.9	60 Nd 144.2	61 Pm 144.9	62 Sm 150.4	63 Eu 152.0	64 Gd 157.3	65 Tb 158.9	66 Dy 162.5	67 Ho 164.9	68 Er 167.3	69 Tm 168.9	70 Yb 173.0	71 Lu 175.0

Actinides 7:

90 Th 232.0	91 Pa 231.0	92 U 238.0	93 Np 237.0	94 Pu 239.1	95 Am 241.1	96 Cm 244.1	97 Bk 249.1	98 Cf 252.1	99 Es 257.1	100 Fm 257.1	101 Md 258.1	102 No 259.1	103 Lr 262.1

Answers

Answers to Odd-Numbered Exercises and Problems

Solutions to questions posed in the Part Overview captions can be found at the end of this answer list.

Chapter 20

1.
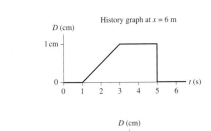
History graph at $x = 6$ m

3.
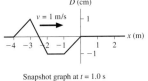
Snapshot graph at $t = 1.0$ s

5.

7. 283 m/s
9. 140 m/s
11. a. 4.19 m b. 47.7 Hz
13. a. 11.5 Hz b. 1.14 m c. 13.1 m/s
15. 4.0 cm, 12 m, 2.0 Hz
17. 40 cm
19. 34 Hz, 68 Hz
21. 0.076 s
23. 793 m
25. a. 1715 Hz b. 1.50 GHz c. 987 nm
27. a. 10 GHz b. 0.167 ms
29. a. 1.50×10^{-11} s b. 3.38 mm
31. 459 nm
33. 6.05×10^5 J
35. a. 1.11×10^{-3} W/m^2 b. 1.11×10^{-7} J
37. 38.1 m/s
39. a. 432 Hz b. 429 Hz

41. a.
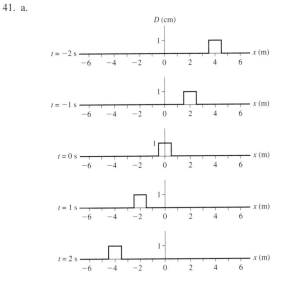

b. 2 m/s c. 2 m/s
43. a. 0.80 m b. $-\frac{1}{2}\pi$ rad
    c. $D(x, t) = (2.0 \text{ mm})\sin(2.5\pi x - 10\pi t - \frac{1}{2}\pi)$
45. 25 g
47. 2.34 m, 1.66 m
49.
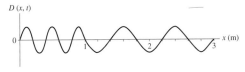

51. 1, 4.31, 4.31
53. 0.07°C
55. a. $-x$ direction b. 12.0 m/s, 5.0 Hz, 2.62 rad/m c. -1.50 cm
57. $D(y, t) = (5.0 \text{ cm})\sin[(4\pi \text{ rad/m})y + (16\pi \text{ rad/s})t]$
61. $\pi/2$ rad $= 90°$
63. a.
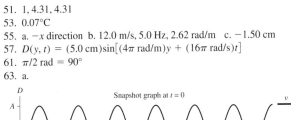
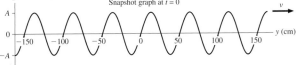
Snapshot graph at $t = 0$

b. 0 rad c. $D(y, t) = A\sin[(12.57\text{ m}^{-1})y - (4310\text{ s}^{-1})t]$
d. and e.

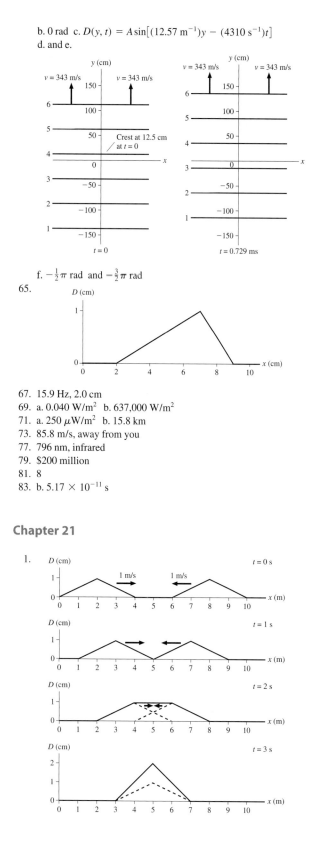

f. $-\frac{1}{2}\pi$ rad and $-\frac{3}{2}\pi$ rad

65.

67. 15.9 Hz, 2.0 cm
69. a. 0.040 W/m² b. 637,000 W/m²
71. a. 250 μW/m² b. 15.8 km
73. 85.8 m/s, away from you
77. 796 nm, infrared
79. $200 million
81. 8
83. b. 5.17×10^{-11} s

Chapter 21

1.

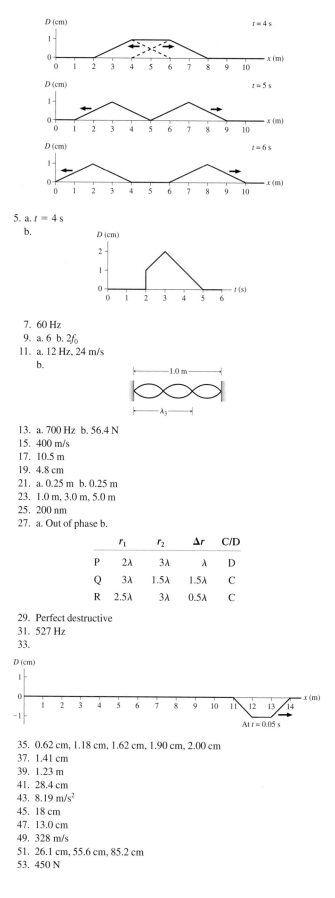

5. a. $t = 4$ s
   b.

7. 60 Hz
9. a. 6 b. $2f_0$
11. a. 12 Hz, 24 m/s
    b.

13. a. 700 Hz b. 56.4 N
15. 400 m/s
17. 10.5 m
19. 4.8 cm
21. a. 0.25 m b. 0.25 m
23. 1.0 m, 3.0 m, 5.0 m
25. 200 nm
27. a. Out of phase b.

	r_1	r_2	Δr	C/D
P	2λ	3λ	λ	D
Q	3λ	1.5λ	1.5λ	C
R	2.5λ	3λ	0.5λ	C

29. Perfect destructive
31. 527 Hz
33.

35. 0.62 cm, 1.18 cm, 1.62 cm, 1.90 cm, 2.00 cm
37. 1.41 cm
39. 1.23 m
41. 28.4 cm
43. 8.19 m/s²
45. 18 cm
47. 13.0 cm
49. 328 m/s
51. 26.1 cm, 55.6 cm, 85.2 cm
53. 450 N

55. 1605 Hz
59. 7.89 cm
61. a. 850 Hz b. $-\frac{1}{2}\pi$ rad
63. 345 nm
65. 7.15 cm
67. 20
69. a. 170 Hz b. 510 Hz and 850 Hz
71. 150 MHz
73. a. a b. 1.0 m c. 9
75. a. 5 beats/s b. 4.6 mm
77. 7.0 m/s
79. b. 2.0%
81. 8.00 m/s^2
83. c. 2.09 cm/s d. 2.2 mm
85. a. $\frac{1}{4}\lambda$ b. $\frac{1}{2}\pi$ rad c. $\frac{1}{4}T$ d. 75 m, 250 ns

Chapter 22

1. a.

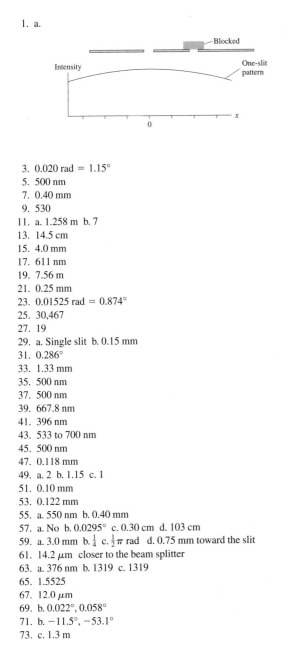

3. 0.020 rad = 1.15°
5. 500 nm
7. 0.40 mm
9. 530
11. a. 1.258 m b. 7
13. 14.5 cm
15. 4.0 mm
17. 611 nm
19. 7.56 m
21. 0.25 mm
23. 0.01525 rad = 0.874°
25. 30,467
27. 19
29. a. Single slit b. 0.15 mm
31. 0.286°
33. 1.33 mm
35. 500 nm
37. 500 nm
39. 667.8 nm
41. 396 nm
43. 533 to 700 nm
45. 500 nm
47. 0.118 mm
49. a. 2 b. 1.15 c. 1
51. 0.10 mm
53. 0.122 mm
55. a. 550 nm b. 0.40 mm
57. a. No b. 0.0295° c. 0.30 cm d. 103 cm
59. a. 3.0 mm b. $\frac{1}{4}$ c. $\frac{1}{2}\pi$ rad d. 0.75 mm toward the slit
61. 14.2 μm closer to the beam splitter
63. a. 376 nm b. 1319 c. 1319
65. 1.5525
67. 12.0 μm
69. b. 0.022°, 0.058°
71. b. $-11.5°$, $-53.1°$
73. c. 1.3 m

Chapter 23

1. a. 3.33 ns b. 0.75 m, 0.67 m, 0.51 m
3. 8.0 cm
5. 668 m
7. 9.0 cm
9. 42°
11. 433 cm
13. 65.0°
15. 1.37
17. 76.7°
19. 3.18 cm
21. 1.52
23. b. 1.1°
25. 1580 nm
27. Inverted image 15 cm behind the lens
29. Upright image 6 cm in front of the lens
31. 68 cm
33. -203 cm
35. 1.54 cm
37. 54.6 km
39. b. Relative to the intersection of the two mirrors, 3 images are at coordinates $(+1$ m, -2 m$)$, $(-1$ m, $+2$ m$)$, and $(+1$ m, $+2$ m$)$
    c.

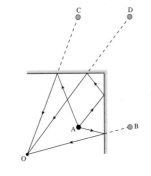

41. 10.0 m
43. 41.8°
45. 82.8°
47. a. Bottom of tank coming up b. 60.0 cm
49. 4.73 m
51. a. Deep b. 17.5 m
53. 1.552
55. a. 17.9° b. 27.9° to the left of the normal
57. 3.0 cm
59. b. 40 cm, 2 cm
61. b. -60 cm, 8.0 cm
63. b. -8.6 cm, 1.14 cm
65. 44.4 cm, 67 cm
67. c. ≈ 3.6 cm
69. 15.7 cm
71. b. 20 cm in front of second lens, 2.0 cm tall
73. 11.5 km
75. a. 2 μm b. 165 MB
77. b. 1.574
79. b. 40 cm, 156.5 cm
81. a. -200 cm

Chapter 24

1. 410.3 nm, 389.0 nm, 379.9 nm
3. $n = 8$
5. 63.8°
7. 4
9. 1.99×10^{-16} J

11. 1.2×10^5 J
13. a. 3.6×10^6 m/s b. 2.0×10^3 m/s
15. a. 1.1×10^{-34} m b. 1.7×10^{-23} m/s
17. 0.20 nm
19. a. 121.6 nm, 102.6 nm, 97.3 nm, 95.0 nm b. 91.18 nm c. 31.4 cm
21. a. 2.0×10^{-12} m b. 2.51×10^5
23. a. 3.14×10^{-19} J b. 3.19×10^{15}
25. 18.7°, 50.8°, and 71.6°
27. b. 2.4 nm and 1.2 nm
29. a. 0.818 μm b. 1.09×10^3 m/s
31. 170 m/s
33. a. 1.23×10^{-19} J, 4.92×10^{-19} J, 1.11×10^{-18} J
    b. 3.69×10^{-19} J c. 539 nm
35. 1.35 nm
37. a. $(h/2mL)n$ b. 1.819×10^6 m/s, 3.64×10^6 m/s, 5.46×10^6 m/s
39. a. 72.5°, 53.1°, and 25.8° b. 64.9° and 31.9°
    c. 19.9° and 76.9°, matching the peaks in Figure 24.7c
41. b. 7.28×10^4 m/s, 1.46×10^5 m/s, 2.18×10^5 m/s, 2.91×10^5 m/s

Part Overview Solutions

PART V Overview

If violent storms drive waves across the ocean of planet Kamino at a speed of 75 m/s, what is the wavelength of an ocean wave whose period is 10 s?

MODEL Assume the waves are sinusoidal with period $T = 10$ s.

SOLVE The fundamental relationship of sinusoidal waves is $v = \lambda f$. The frequency is related to the period by $f = 1/T$. Consequently, the wavelength is

$$\lambda = vT = (75 \text{ m/s})(10 \text{ s}) = 750 \text{ m}$$

ASSESS 750 is much larger than the wavelength of ocean waves on earth because the 75 m/s $\approx$ 150 mph wave speed on Kamino is much larger than the wave speed on earth.

Credits

All Part Overview images are courtesy of Lucasfilm Ltd. Addison Wesley would like to give special thanks to Lucy Wilson, Christopher Holm, and the staff of Lucasfilm Ltd. for granting us permission to use these images and for their help in selecting them.

TITLE PAGE
Page **iii:** Rainbow/PictureQuest.

PART V
Part V Overview image: *Star Wars: Episode II – Attack of the Clones* © 2002 Lucasfilm Ltd. & ™. All rights reserved. Used under authorization. Unauthorized duplication is a violation of applicable law.

CHAPTER 20
Page **611:** Reuters NewMedia Inc./Corbis. Page **612** B: Doug Wilson/Corbis. Page **612** T: Hoard Dratch/The Image Works. Page **614:** Uri Haber-Schaim. Page **629:** V.C.L./Getty Images. Page **631:** David Parker/Photo Researchers, Inc. Page **633:** Doug Sokell/Visuals Unlimited. Page **638:** Space Telescope Science Institute.

CHAPTER 21
Page **646:** Rosco Permacolor™ filter: Rosco Laboratories, Inc. Page **648:** Richard Megna/Fundamental Photographs. Page **652:** Education Development Center, Newton, MA. Page **654:** Tom Pantages. Page **658:** Richard Gross/Corbis. Page **659:** Uri Haber-Schaim. Page **665:** Peter Aprahamian/Photo Researchers.

CHAPTER 22
Page **684:** Chris Collins/Corbis. Page **685** T: Richard Megna/Fundamental Photographs. Page **685** B: Todd Gipstein/Getty Images. Page **686:** Springer-Verlag GmbH & Co KG. Page **687:** M. Cagnet et al., Atlas of Optical Phenomena Springer-Verlag,1962. Page **692:** M. Cagnet et al., Atlas of Optical Phenomena Springer-Verlag,1962. Page **693:** Courtesy Holographix LLC. Page **699** T, B: Ken Kay/Fundamental Photographs. Page **700:** Springer-Verlag GmbH & Co KG. Page **704:** CENCO Physics/Fundamental Photographs. Page **706:** Dr. Rod Nave, Georgia State University.

CHAPTER 23
Page **714:** Charles O'Rear/Corbis. Page **718:** Sylvester Allred/Fundamental Photographs. Page **721:** Richard Megna/Fundamental Photographs. Page **725:** Richard Megna/Fundamental Photographs. Page **726:** Francisco Cruz/SuperStock. Page **730:** Tony Freeman/PhotoEdit. Page **731:** Benjamin Rondel/Corbis. Page **732** L, R: Richard Megna/Fundamental Photographs. Page **736:** Photodisc Blue/Getty Images. Page **747:** Benjamin Rondel/Corbis. Page **748** L, M, R: Springer-Verlag, GmbH & Co KG.

CHAPTER 24
Page **757:** IBM/Phototake NYC. Page **758:** Courtesy of Ocean Optics, Inc. Page **759:** Gerard Herzberg/Atomic Spectra and Atomic Structure, Prentice-Hall,1937. Page **762:** General Electric Corporate Research & Development Center. Page **763:** General Electric. Page **763** L, M, R: Eugene Hecht/Optics 2e, p.11, Addison Wesley. Page **764** T, M, B: E.R. Huggins, Physics I/W.A. Benjamin, 1968, Reading, MA, Addison Wesley Longman, reprinted with permission. Page **769** L, M: Film Studio/Education Development Center, Newton, MA. Page **769** R: Tipler and Llewellyn/Modern Physics 3e, 1987, p. 207, New York, NY, Worth Publishers, courtesy of C.G. Shull. Page **769:** Dr. Claus Jonsson. Page **770:** P. Merli and G. Missiroli, *American Journal of Physics*, 44, p. 306, 1976. Page **774:** P. Merli and G. Missiroli, *American Journal of Physics*, 44, p. 306, 1976.

Index

For users of the five-volume edition: pages 1–481 are in Volume 1; pages 482–607 are in Volume 2; pages 608–779 are in Volume 3; pages 780–1194 are in Volume 4; and pages 1148–1383 are in Volume 5.
Pages 1195–1383 are not in the Standard Edition of this textbook.